'This is an absolute winner: entertaining, educational, jaw dropping. Never before have I been so captivated by a piece of non-fiction. There are various good-enough books out there, but this is something else.' *Bookbag*

'Blind optimism is the last thing the world needs – a recipe for disappointment. But what about optimism based on careful reasoning? On digging below the surface for the ideas and the trends that really do add up to something promising? That's what's on offer here. Stevenson wears no blindfold. His tools are curiosity, open-mindedness, clarity and reason. That makes his journey intriguing ... and exhilarating.' Chris Anderson, TED Curator

'Uplifting and liberating. Stevenson is optimistic without being naive, fun but not silly, entertaining yet also an educator. By the end I was utterly convinced that, given the opportunity to flourish, human ingenuity can disperse the clouds that hang over us today. This is a book to gladden the gloomiest heart.' Michael Brooks, author of *13 Things That Don't Make Sense*

'Makes a good case for believing that we can have a future worth making an effort to reach.' Jon Turney, *Guardian*

'Infectiously enthusiastic' James McConnachie, *Sunday Times*

'Refreshingly upbeat' *Big Issue*

'The future cannot happen until we change our minds to meet it' *Focus*

'Essential ... illuminating and refreshingly hopeful ... an auspicious yet grounded vision' *The Atlantic*

'Stevenson describes our future's possibilities with a journalist's eye for detail, a teacher's knack for translating complexities, and a comic's wry commentary.' *Chr*

'An ability to express even the most complex scientific problems in terms easily understood by a layperson.' *Sydney Morning Herald*

'Resource crises, climate change, terrorism, the advancement of machine technology, it's all here – complete with references to Burroughs, *The Hitchhiker's Guide to the Galaxy*, Proust and Einstein. In questioning progress, Stevenson employs science, ethics and philosophy and never fails to be inspirational.' *Mind Food Magazine*

'*An Optimist's Tour of the Future* may just be our generation's version of Bill Brysons's iconic *A Short History of Nearly Everything* – a bold and entertaining blueprint for a future that's ours to shape and ours to live.' *Brainpickings*

'This is a brilliant book, and Mark Stevenson is the perfect guide to a dizzying future that is already here. The ideas come so quickly, with such great humour – it's like the smartest dinner party you've ever attended.' Peter Miller, author of *The Smart Swarm*

MARK STEVENSON divides his time between running agencies for science communications and cultural learning and performing stand-up comedy. He lives in Telegraph Hill, south London.

An Optimist's Tour of the Future

Mark Stevenson

P
PROFILE BOOKS

This updated paperback edition published in 2012

First published in Great Britain in 2011 by
PROFILE BOOKS LTD
3A Exmouth House
Pine Street
London EC1R 0JH
www.profilebooks.com

10 9 8 7 6 5 4 3 2 1

Set in Adobe Jenson Pro and Mr Eaves to a design by Henry Iles

Printed and bound in Great Britain by
CPI Group (UK) Ltd, Croydon, CR0 4YY

A CIP catalogue record for this book is available from the British Library.

ISBN 978 1 84668 357 2
eISBN 978 1 84765 433 5

For Charlie (no. 23) and Philip (no. 55).

The future is here.
It's just not widely
distributed yet.

WILLIAM GIBSON

Contents

PART 1

MAN

CHAPTER 1

The world's most dangerous idea

Getting older is no problem. You just have to live long enough. GROUCHO MARX

I'm on a train to Oxford wondering how long I am going to live. This isn't because I think we're about to crash, nor is it a reaction to the sandwich First Great Western has just sold me. It's because a few weeks ago the stalking horse of Mortality popped into my head and, without invitation, asked, 'What are you going to do with the rest of your life, then?' (Mortality, it turns out, sounds a lot like my father.)

Whatever the answer, one thing I know is that my future will play out in a very different world. Because – let's be frank – there is a revolution going on. The population is rocketing, the planet is becoming urban (over half of us now live in cities), medicine is curing the previously incurable, ninety-year-olds take parachute jumps, spaceships are owned by businessmen, the climate is shifting and the world's knowledge is becoming available to anyone with an Internet connection.

So I'm setting out on a journey that I hope will tell me what the world of my future will look like. Are we going to cure cancer? What is the 'biotech revolution'? Are the robots finally coming? What's all that nanotechnology stuff about? How will the Internet continue to shape society? How do we handle a growing population? If the climate is changing, how will that affect us and can we do anything about it? Are we going back to space? Is technology going to be our friend or run off into the distance, leaving us dazed and confused? And how will the answers to all these questions tangle together and change the way we live, work and play?

I want to find an answer to the most personal and yet the biggest of questions, a question all of us ask: 'What's next?' But first I need to work out how much 'next' I'm likely to see. Or, to put it another way – how long have I got? How long is my future?

The UK Office for National Statistics states that average life expectancy at birth for a man born in Great Britain in 1971 (that's me) was just over sixty-nine years. It also informs me that, as a thirty-nine-year-old male in 2010, I can now expect to live ten years longer. The mere act of staying alive, it seems, has given me more living to do.

'Life expectancy creep' is well documented. Average age at death has been steadily edging upward by roughly a quarter of a year for every calendar year that's passed since vaguely reliable records began (around the mid-nineteenth century). Some say this trend is likely to continue, even accelerate, in tandem with medical advances, and that centenarians will soon become commonplace. Others argue that the rise in statistical life expectancy is due more to rescuing the young (by fighting infant tuberculosis, for example). Hinting at a natural limit to our longevity, Stuart Jay Olshansky of the University of Illinois School of Public Health points out, 'You can't save the young twice.'

With a few notable exceptions – Ozzy Osbourne, Keith Richards, all of Aerosmith – there is also a well-established link between the way someone lives and their expected life span. Accordingly I've answered a variety of lifestyle and family history questions at the request of several Internet-based life expectancy calculators. My favourite, 'Death Clock,' produces a personal 'death date' based on 'normal,' 'optimistic,' 'pessimistic' or 'sadistic' calculations (the latter announcing you're already dead) before displaying the number of seconds you have left to live ... and then *counting down*. There's something ironically mesmerising about literally watching your life tick away. I wonder if anyone has actually died doing it?

The consensus is that, if I maintain my current lifestyle, I'll live past eighty but not eighty-five (so, roughly in line with government statistics). However, if I improve my diet, get more exercise, continue working (according to some studies, a key strategy for keeping the grim reaper at bay) and drink less, I've a good chance of pushing past ninety. And if life expectancies keep creeping up the way they have been, fifty years from now I could have up to *another* twelve years on the planet. It's not infeasible I'll make it past a hundred. I find this encouraging: I'm still single with no kids, but it appears there's plenty of time to raise a family, maybe even understand cricket.

Nick Bostrom, founder of Oxford University's Future of Humanity Institute, agrees. In fact, he thinks I could live not for a hundred years but for thousands. And he's not joking.

Bostrom is an advocate for transhumanism. Described as 'the world's most dangerous idea,' the concept got its modern name in *Religion Without Revelation*, written in 1927 by Julian Huxley. The brother of Aldous, Julian Huxley was a leading biologist, the first director-general of UNESCO and founder of the World Wildlife Fund. Busy chap. He wrote:

> The human species can, if it wishes, transcend itself – not just sporadically, an individual here in one way, an individual there in another way – but in its entirety, as humanity. We need a name for this new belief.

The term Huxley came up with was 'Transhumanism' – the idea that man could remain essentially human but transcend what nature had given him. Skip forward seventy-seven years to 2004 and you get Aubrey de Grey, a biogerontologist who declares, 'I think the first person to live to a thousand might be sixty already.'

I'm travelling to see Nick because the debate around transhumanism is a car-crash of science, ethics and social policy. It asks us to rethink what medicine is, what society might look like and how far we should go with our technology. I'm kind of throwing myself in at the deep-end.

Via a stuttering wi-fi connection, I use part of my journey to look again at a typical polemic given by de Grey at the 2004 'Technology, Entertainment and Design' (TED) conference in California. In it he asks if anyone in his audience 'is in favour of malaria?' When no one raises their hand he continues, 'I put it to you that the main reason we think malaria is a bad thing is because of a characteristic it shares with ageing – and here is that characteristic' – at which point he shows a slide bearing the words 'Because it kills people!!!!!!' 'The only real difference is that ageing kills considerably *more* people.' De Grey proceeds to compare ageing to fox hunting, claiming they are both 'traditional,' 'keep the numbers down' and, his punch line, 'fundamentally barbaric.' At the same conference, Nick Bostrom later stated, 'Death is a big problem. If you look at the statistics the odds are not very favourable. So far most people who have lived have also died.'

But banishing ageing is only one half of the transhumanists' truly radical agenda. They also advocate a new vision of the human being: the transhuman – a human not only saved from ageing, but *enhanced* beyond our current 'biological limitations.'

As a child I used to play with a transhuman called Steve Austin. Steve was a toy doll of *The Six Million Dollar Man,* a merchandising spin-off from the TV show of the same name. Just like TV Steve, my doll had 'bionic' body parts that helped him see further, run faster and punch harder than his (frequently moustachioed) adversaries. His manufactured body parts were replacements for those lost in a plane crash. Over the opening credits his backstory was intoned: 'We have the capability to make the world's first bionic man ... Better, stronger, faster.'

These last words reflect much of the transhumanists' enhancement credo – that transhumanism isn't only about keeping us alive and healthier for longer, but allowing us to *be more*. Be better. Be stronger. Be faster.

History tells us of numerous individuals who've searched for the secret of living for ever. Chinese alchemist Ge Hong (born in 283 CE) believed immortality was achieved by cultivating everlasting union with yourself, allowing you to sip from a wellspring of a 'metaphysical oneness' or *xuan*, which permeated all things. It's a dream that is just as prevalent today. In San Francisco, for example, Alex Chui will sell you a pair of magnetic finger rings or 'eternal life devices' for around thirty dollars. They come with a ninety-day refund, which seems an extraordinarily unconfident guarantee. When I contact Chui to raise this point he says, without irony, that 'a lifetime guarantee is illegal. You cannot make such a claim.'

Perhaps the best known 'scientific' method of trying to avoid the great gig in the sky is Cryonic Suspension – the process of preserving the recently deceased by freezing their body or, for those with less funds (and I find this deliciously macabre), just their head. This method rests on the belief that physicians in the future will be able to thaw out, re-animate and cure these 'cryonauts' of whatever it was that killed them in the first place.

Keeping someone frozen (or, in the parlance of wittier cryonics enthusiasts, 'metabolically challenged') is far from cheap – currently, you won't get much change out of $150,000. Two early cryonics firms went bankrupt and their customers defrosted as a result, but I suspect none of them asked for a refund. Another problem for early cryonauts is the damage done to their cells by ice crystals forming in tissue, making it highly unlikely that pioneer corpses like psychology professor James Bedford (the first man to be suspended in 1967) will ever be successfully re-animated. Today the cryonics profession has put its faith in 'vitrification' – a technique that converts tissue to a glass-like solid, but free from any crystalline structure. It's fair to say that cryonics hasn't been accepted as a mainstream medical practice and is generally seen as something that wacky people do when they've earned too much money and/or want to annoy their immediate descendants. As far as today's medicine is concerned death is a one-way trip.

I mention these examples because stories of misguided optimism, wishful thinking, moral crusading, faith in unproven technology and, let's be honest here, out-and-out fruitcake-ism are not uncommon in the history of man's quest to cheat death. Accordingly, I'm approaching the transhumanist project with a good dose of scepticism.

I'm not the only one to be dubious. In 2006, fed up with what they perceived as a lot of media 'bad science' stirred up by the anti-ageing part of the transhumanist project, the *MIT Technology Review* offered a $10,000 prize to anyone who could demonstrate that Aubrey de Grey's ideas and research were 'unworthy of learned debate'. Jason Pontin, the *Technology Review* editor had previously written a stinging profile of de Grey, characterising him as a shabby, beer-addicted troll and comparing his work to 'science fiction'. Not one to take offence, de Grey offered a further $10,000 prize money.

The submission that attracted the most praise (written by Preston Estep and colleagues) claimed that de Grey's work enjoys notoriety 'due almost entirely to its emotional appeal' and dismissed much of it as 'pseudoscience'. But the judges (who included world experts in artificial intelligence, nanotechnology, pathology and genetics) didn't think Estep and his colleagues had done enough. While de Grey's ideas exist 'in a kind of antechamber of science, where they wait (possibly in vain) for independent verification' they were 'not demonstrably wrong'. Judge Nathan Myhrvold wrote that 'it is a hallmark of the scientific process that it is fair about considering new propositions; every now and then, radical ideas turn out to be true. Indeed, these exceptions are often the most momentous discoveries in science.'

History is, of course, littered with mavericks initially ridiculed or dismissed and later (sometimes *much* later) accepted as geniuses: including Charles Darwin (evolution), Gregor Mendel (genetic inheritance), Robert Goddard (liquid-fuelled rockets),

Louis Pasteur (germ theory) and the Wright brothers (powered flight). Then again, the past is also full of challenges to accepted wisdom which crashed and burned. I give you James McConnell and Georges Ungar, who believed that memories were encoded in molecules and could therefore be transferred from one animal to another – giving rise to the possibility that you could take a pill and go on to recall the complete works of Shakespeare.

In summary, Bostrom and de Grey *might* be on to something. And if transhumanists really are the outriders of the future, that, as a neuroscientist friend of mine observed, 'is a bit of a headfuck.'

Which is another way of saying that Transhumanism raises deep philosophical and social questions. What will the impact on society and personal relationships be in a world where we might live for ever? Would any of us get married if we knew it was for a thousand years? When do we retire? Isn't there a massive overpopulation problem? Won't we run out of natural resources? If we merge with our machines, what might we lose? What about pensions? What will *Saga* magazine do? And isn't death natural, part of the cycle by which our species evolves? Isn't ageing a defining characteristic of being human? Wouldn't we go mad? And even if we could defeat death, should we?

With middle age at my back the idea of being rejuvenated to a younger body is very attractive. But I worry that what I think might be good for me might not be good for humankind as a whole. To their credit, most transhumanists accept that ending ageing (if it's possible) won't arrive without massive upheavals to society. The question Aubrey de Grey asks is, '[Do the risks] outweigh the downside of doing the opposite, namely leaving ageing as it is ... condemning a hundred thousand people a day to an unnecessarily early death? You know, if you haven't got an argument that's that strong then just don't waste my time.' For transhumanists this is a moral argument.

On the issue of over-population Aubrey de Grey has written, 'Basically our options are extremely simple: either restrict the birth rate or raise the death rate,' by which he means if we all start living longer and keep to one planet, we'll probably have to breed less. Bostrom is more circumspect. Such questions, he has stated, should 'be addressed in a sober, disinterested way, using critical reason and our best available scientific evidence' in order to decide 'what policies it makes sense for humanity to pursue.' I admire his sentiments, but I wonder if asking us to think about all this rationally is like asking a fly to consider the aesthetic qualities of a windshield.

Still, this is why I've come to Oxford ...

In the same week that I visit Professor Bostrom, two news stories catch my eye. The first concerns scientists at the Spanish National Cancer Centre in Madrid who have been breeding mice which live, on average, *thirty per cent longer* than their 'normal' brethren. Led by Maria Blasco, the team is building on the work of Elizabeth Blackburn, Carol Greider and Jack Szostak, who discovered a startling substance (strictly an enzyme – a substance cells use to spur on chemical reactions) called 'telomerase' while they were investigating a micro-critter named 'Tetrahymena' found in freshwater ponds, a discovery for which the three shared the 2009 Nobel Prize for Medicine.

Tetrahymena has two particularly interesting characteristics. One, it has *seven* sexes. Two, it's biologically immortal. You read that right. If it can eat, it just keeps on living – and telomerase is one of the reasons why. For Tetrahymena isn't subject to something called the 'Hayflick limit,' named after medical microbiologist Leonard Hayflick.

Along with his colleague Dr Paul Moorhead, Hayflick worked out that human cells have a built-in 'age-then-die' program that kicks in after they've divided somewhere between forty and sixty times. When the cell reaches this limit, it enters its last days – a state called 'senescence.' (De Grey calls his assault on ageing 'Strategies for Engineered Negligible Senescence.') But a constant supply of telomerase (which Tetrahymena continually replenish themselves with) can keep resetting a cell's 'divide count.' This means that even if a cell is fifty generations down the line, it behaves like it's only divided once (keeping it well away from the death knell of the Hayflick limit).

Human telomerase levels decline as we get older, so the natural question for the anti-ageing brigade is, 'If we increase our supply of telomerase, can we combat ageing?' Several studies have shown that people who live healthy lifestyles (and therefore tend to live longer) have higher levels of telomerase in certain cell lines, supporting a direct link between longevity and telomerase – and increasing the supply of it to Blasco's mice did indeed dramatically lengthen their lives. Against the constant backdrop of continuing medical advances, stories like this seem to suggest that the transhumanists may not be quite as crazy as they might first appear.

But there's always a catch. Cancer cells also come with a helpful supply of telomerase – which is one of the reasons they divide and multiply with such awful success. The telomerase helps keep them from entering senescence too. Cancer, after all, is defined as the uncontrollable division of cells. (That's what tumours are – cancerous cells that have gone on a dividing frenzy without quitting.) So, if you go for a telomerase top-up, like Blasco's mice, you'd need to be bred cancer-resistant too.

The second story to grab my attention that week is more compelling. It was about a woman called Claudia Castillo, who in 2008 suffered a collapse of part of the windpipe leading to her left lung. The result was a debilitating shortness of breath, making it impossible for her to look after her children, to work, clean, shop or cook.

The usual medical options for such a condition are wholesale removal of a lung or a partial trachea (windpipe) transplant. Transplants are particularly tricky because our immune systems tend to reject donated body parts, requiring patients to take immunosuppressive drugs. These reduce your body's desire to reject the new part, but at the same time can severely hamper its ability to combat infection. Instead, aged thirty, Castillo became a pioneer for 'stem cell' therapy – one of the most rapidly advancing areas of medical science.

Stem cells are cells that are yet to pick a career – or to put it another way they are 'blank' cells, which can ultimately be 'programmed to perform particular tasks.' This is why human embryos naturally have a ready supply and is where researchers have often got them from (leading to all that controversy). But adults also carry around their own supply of 'tissue-specific' stem cells. These cells are also still waiting to be 'programmed' but have a smaller palette of career choices, based on their tissue of origin. That makes them less controversial (no baby parts) but less widely applicable (they can only turn into particular bits of you). It's now possible, however, to take ordinary cells (say, from the skin on your hand) and 're-boot' them back to their earlier stem cell selves, doing away with many ethical concerns altogether. But even that may not be necessary. Early in 2010, scientists at Stanford University in California announced they had managed to convert mouse skin cells *directly* into mouse brain cells (with no intermediate stem cell phase). A few months later a team from Harvard declared they could convert human blood cells to stem cells that may have the ability to grow into any kind of tissue.

Castillo received a donated trachea that had been stripped of cells from the donor (leaving just a collagen pipe). This was then 'seeded' with stem cells harvested from her own body (and nurtured by a team at the University of Bristol) that grew into tissue. The result? Her body accepted the new windpipe as its own and medical history was made. It sounds like science fiction, but the patient is out dancing.*

Castillo isn't the only person wandering around with a stem cell-grown body part. Kaitlyne McNamara's doctors isolated healthy adult stem cells from the diseased bladder she was born

* An interesting coda to Claudia's story is that she nearly didn't get her operation. With a time limit of sixteen hours for Claudia's lab-nurtured cells to reach their destination, the Bristol team booked the only direct Bristol-to-Barcelona flight – operated by EasyJet. Despite getting agreement from EasyJet several months in advance that student Philip Jungerbluth could carry Claudia's cells as hand luggage, check-in staff flatly refused to allow the material onto the plane, citing 'anti-terror' security measures. Professor Martin Birchall was nearly arrested by armed police as he fought vainly against the orange-uniformed face of low-cost air travel. The answer? A friend of Philip's (a pilot) and €11,000 out of Birchall's own pocket to charter a private flight.

with and used them to grow an entire, fully functioning (and healthy) replacement, which they then implanted. The Wake Forest Institute for Regenerative Medicine in Winton-Salem, North Carolina are working on more than twenty different organs and tissues including kidneys, livers, retinas and muscle. Doris Taylor, Harold Ott and colleagues at the Centre for Cardiovascular Repair at the University of Minnesota took the collagen shell of a rat's heart, sprayed it with stem cells and it started beating. 'When we saw the first contractions we were speechless,' says Ott. In February 2010 the technology was licensed to a company called Miromatrix Medical, which hopes to revolutionise organ transplants with the technique.

There are three fronts on the transhumanists' war with our inbuilt human limitations. The first is an attack on the restrictions of our biology. If the transhumanists wanted a field marshal to lead this advance, then Aubrey de Grey would be the obvious candidate. He's encouraging the scientific community to find solutions to 'seven major types of molecular and cellular damage that eventually become bad for us' which he believes, once fixed, will put an end to ageing. He's a big fan of stem cells and long-lived mice (his Methuselah Foundation offers prizes for scientific research teams who extend mouse life spans). De Grey's line of attack could be called the 'wet front' – dealing as it does with the slippery world of cells and their biology.

The second front could be called the 'dry front' – evolving from our experiences building machines. Here the transhumanists could reasonably propose Ray Kurzweil (a man I add to my list of those I should meet later on my journey) as the commander in chief – one of the world's key thinkers in the area of artificial, or machine, intelligence as well as being a serial entrepreneur and genius.

Kurzweil believes we're likely to see smarter-than-human machine intelligences well before the end of the twenty-first century. He also believes that the line between humans and machines will blur as we use machine technologies to enhance our abilities, give our brains cognitive boosts and, among other things, upload the human consciousness to a hard disc. Ray talks of a 'merger of our biological thinking and the existence of our technology, resulting in a world that is still human but transcends our biological roots.' This merger 'will allow us to overcome age-old human problems and vastly amplify human creativity. We will preserve and enhance the intelligence that evolution has bestowed on us while overcoming the profound limitations of biological evolution.'

The third front in the war against human limitation is perhaps the most important of all – the 'what the hell does it all mean and what should we do about it?' front. And if the transhumanists were looking for an admiral of the fleet, then Professor Bostrom would certainly be a front runner. This is his self-penned biography:

> Beside philosophy, I also have a background in physics, computational neuroscience, mathematical logic, and artificial intelligence. My performance as an undergraduate set a national record in Sweden. I was a busy young man. Before becoming a tweedy academic, I also dabbled in painting and poetry, and for a while I did stand-up comedy in London.

He is thirty-six. I'd hate to read his biography when he's fifty, by which time he will no doubt have enjoyed a brief stint in the West End as a dancer, proved a number of the unsolved problems of mathematics, built a robot wife and published several successful cookbooks.

Arriving at Oxford I head for Broad Street and the Indian Institute, where Professor Bostrom has invited me to an afternoon seminar discussing transhumanist ethics. Inside I find a modern, plastic-chaired seminar room (a sharp contrast to the stonework Hindu demigods and tigers' heads outside). I'm the first to arrive but over the next fifteen minutes a cadre of serious-browed, sensibly-shoed, studious-looking types drift in. Bostrom is one of the last to turn up, just in time for the first presentation, given by Rob Sparrow.

Sparrow is an academic I instantly warm to, an Australian bioethicist with a cheeky look in his eye, who's not ashamed to say that he is 'genuinely confused' by the implications of his own work. Rob's worried that some thinkers in the transhumanist movement are ushering in a world of 'free market eugenics.' He is particularly vexed by the principle of 'procreative beneficence', an idea advocated by Julian Savulescu, an Oxford contemporary of Bostrom's.

The Savulescu argument goes something like this: If after birth you wilfully altered your child's genetic makeup to give her a medical condition that would reduce the quality of her life (say, asthma or Asperger's) this would be considered abuse, akin to beating your offspring. Now consider a couple undergoing IVF (in vitro fertilisation) who must select one of two embryos which are genetically identical except one has a defective gene that is likely to result in asthma. 'Procreative beneficence' argues that if the technology is available to give parents this knowledge, choosing anything other than the 'healthy' embryo is analogous to abuse (although the principle still defends the right of parents to make the 'wrong' choice).

Rob Sparrow doesn't like the implications of this kind of thinking, because I suspect he fears that, even with caveats, we are on dodgy moral ground. The phrase 'thin end of the wedge' comes to mind. He asks us to imagine parents choosing between two embryos – one which has a condition that, on balance, means it will probably live five years less than the other. 'Which embryo should they pick? Embryo A with the condition, Embryo B without the condition or should they choose the third option, chance, and toss a coin?' I am the only person in the room to choose tossing a coin because I figure quality of life matters more than the length of it. It simply turns out that the embryo with the shorter projected life span is a boy.

Soon we're considering the argument that the future is better off all-female. Women live longer and tend to nurture offspring more diligently; they are less violent. It's theoretically possible to manufacture sperm and egg from bone marrow, doing away with the biological necessity for Brad Pitt – and with a little bio-tinkering all women could be created lesbian, doing away with the desire for Pitt as well. But, of course, you could equally well argue that, with human-altering technologies, it might be just as reasonable to reduce men's proclivity to violence, extend their life spans and give them wombs. Sparrow cites the 'gap' some men feel by not being able to bear children (I'm not one of them) and wonders, if there is a demand for male pregnancy, should it be an option freely available if the technology can be made safe?

I suddenly realise that for these assembled academics, the prospect of fundamental alterations of our biology (including radical life extension and human enhancement) is a *given*. They're not talking about what happens *if* it becomes possible, they're discussing what we might do *when* it's an option. And this worries many transhumanist critics. Sparrow later tells me, 'The implications of taking transhumanism seriously are so radical and implausible that I think we should be much less inclined to do so.'

The seminar ends, and as the attendees dissipate (several I notice lighting cigarettes as they leave the building), I introduce myself to my host. Professor Bostrom is tall and slim with fine blond hair falling over a large forehead. His face is angular and inquisitive with deeply set eyes – giving the impression it has spent a good deal of time screwed up in concentration, which on reflection is probably just how the face of a philosopher should look. There's a kindness to his features – and a certain impishness there too. When he talks, he almost chews his words, delivering each with deliberate clarity – the native Swedish inflections giving his English an authoritative 'weight.' As we walk to his office, I ask Nick how many transhumanists like a smoke.

'They fall into two camps,' he observes. 'Some are trying to stay as healthy as possible so they'll live long enough to get anti-ageing therapies.' He pauses. 'Others believe that we're near enough to the technology to enjoy ourselves and not worry about the damage.'

We turn down a nondescript side street and arrive at the Future of Humanity Institute, which is not nearly as grand as its title implies. In fact, the building we stop at could easily be an insurance company sub-branch. The institute itself comprises just four modern offices and a meeting area on the second floor. The professor's office might be that of a loss adjuster.

So, is transhumanism really 'the world's most dangerous idea?', I ask.

'It *is* a dangerous idea,' he asserts. 'Once you start to modify our basic biology, if you go about it *unwisely*, the results could be . . .' He pauses. 'Um, quite negative.' But, he reasonably points out, 'the bigger the potential risks are – of course, the bigger the potential upsides as well. So the key is to try and be wise and ethically responsible. Which is easier said than done.'

He sees a radically different future shaped hugely by the advances of science and medicine. But I also get the impression he feels those advances are not being equalled by similar improvements in our ability to deal with the possible moral and ethical consequences.

Professor Bostrom apparently sees his role as trying to help our moral compass catch up. He explains this by referencing ants and Adolf Hitler. 'Think of us like ants. Maybe you're a happy little ant, building a good ant hill. But maybe you're contributing to Hitler's war machine? I'm arguing it's better to wake up and try and understand the bigger context.'

By his own admission this is something that human beings are 'very bad at' and 'often get wrong.' He even suggests it may be beyond us, but this hasn't stopped him trying. 'I want to think rationally about these things, rather than going for whatever makes the best story or gives you the most emotional comfort. That's a very big challenge.'

It's important to remember that Nick comes from a very pro-transhumanist perspective. He is, after all, the joint founder of the World Transhumanist Association, which 'advocates the ethical use of technology to expand human capacities' supporting 'the development of and access to new technologies that enable everyone to enjoy better minds, better bodies and better lives.'

I'm keen to understand where his optimism about transhumanist outcomes comes from, when so many people I mention the idea to visibly recoil. What drives him to bang the transhumanist drum in a world generally hostile to the idea? Throughout our talk, I pose this question in various guises, and Bostrom always answers in the same spirit – that, in his view, human life extension and enhancement will allow us not simply to live longer, but to *enjoy* living much more. When I ask, 'What inspires and motivates you?' he cites his reading, his colleagues, but also states 'I guess through feeling and experiencing something in this life and thinking "why can't it always be as good as that?"'

This is the emotional driver at the core of the transhumanist dream. 'It's hard for a lot of people to see the problem – that life isn't always as wonderful as it could be – and part of the reason for that is our biology. Just like chimpanzees (or ants) we have biological constraints in the sort of thoughts we can have and the lives we can lead. To overcome this, the transhumanist project is developing technologies that can enable us to become all that we are in potential, by changing not just the world around us but human biology as well.' Well, it's certainly more ambitious than getting a new kitchen. For him and many other transhumanists it's not just 'better than human,' it's also 'have a better time than humans.' And if you're going to have a better time, why stop at seventy or eighty years?

'Life extension is very central to me,' he states. 'First because for all the ordinary reasons we prefer to be alive rather than dead and, second, because if you think that the future might hold additional possibilities of radical enhancements of human capacity, the only way to get to experience that is to remain alive long enough, until those technologies have been developed.'

Remember, when transhumanists talk of enhancement, they're going way beyond the capacity to switch on that fleeting 'two pints in: brilliant at pool' ability at will. Transhumanists talk of boosting your intelligence, giving you Steve Austin's eyesight, access to a perfect memory or the ability to run faster than an Olympian. Again, all good for the individual, but is it good for our species as a whole?

In a paper entitled 'Protecting the Endangered Human,' bioethicists George Annas, Lori Andrews and Rosario Isasi worry that:

> The new species, or 'posthuman,' will likely view the old 'normal' humans as inferior, even savages, and fit for slavery or slaughter. The normals, on the other hand, may see the posthumans as a threat and if they can, may engage in a pre-emptive strike by killing the posthumans before they themselves are killed or enslaved by them.

They go on to argue that the potential for inter-species genocide makes the transhumanist agenda one of potential mass destruction. This is on the apocalyptic end of the criticisms of transhumanism, but it's all too possible to imagine a scenario where, through cost or distribution or political agenda, some people get the opportunity to enhance themselves and lengthen their lives, while others don't. Isn't it very plausible that we could end up in a two-tier society of posthumans and unenhanced 'normal' humans – with huge potential for conflict and disharmony, even war?

Bostrom argues that it doesn't necessarily follow that enhancement technologies will increase inequality. 'So on the one hand people who are richer and better informed will presumably have better access to the latest technologies. But on the other it seems to be technologically easier to enhance the capacities of those who start off on a lower level. It just seems to be easier to fix a system that is already not functioning optimally. The further you are away from this optimum the more you start to benefit. This would tend to *reduce* inequalities, because it's just easier to lift up the floor than it is to raise the ceiling.'

In short, it's easier to help those of us with 'suboptimal' genetics and health than to improve the prospects of those of us lucky to have been already dealt good cards. Looking at the link between income and life expectancy on an international scale and across time, Professor Bostrom's argument appears to be writ large.

In 1901, average life expectancy in the UK was forty-eight, and average income per person per year $6,264 measured in international dollars fixed at 2005 prices. (International dollars are a useful hypothetical unit of currency that measures value in terms of 'purchasing power' – thereby allowing like-for-like comparisons of relative wealth across geographies and timescales.) A hundred and two years later, in 2003, the African state of Burundi achieved the life expectancy rates the UK enjoyed in 1901, but on an average yearly income of just $427 international dollars. Or put another way, Burundians are enjoying the same life expectancies of people who were nearly *fifteen times richer* than they are. It turns out that across the world the difference in longevity between rich and poor has narrowed and is continuing to do so.

I challenge Nick with the opposite scenario – one where there is equal access to enhancement technologies. Is it possible this could slowly eradicate diversity? 'For instance,' I say, 'we're very aware of homophobia. Could we "enhance" somebody to become straight? So, rather than fight prejudice, why not just have a "straightening" therapy?'

Then Bostrom says something that turns a key inside me. 'I think we need to distinguish between good and bad forms of diversity. If nobody had lung cancer, there would be slightly less difference in the world. That's less diversity. But most people think it would be a good thing if we had less cancer.'

This is my lightbulb moment when I 'get' the transhumanist view of the future and why they think it's inevitable. Ask someone 'Do you think we should live for ever?' and they'll probably say no. Ask the same person if we should continue to battle disease, they'll almost certainly say yes, me included. Transhumanism won't arrive in a revolution, it'll arrive one therapy at a time. Fading eyesight? Try large print. Now these spectacles. Now contact lenses. Now laser eye surgery. Now stem cell therapy to return your eye to its former youth. (In fact Italian researchers have been using stem cells to cure corneal blindness for over ten years.) Bad hearing? Have a large cone to stick in your ear. Now a bulky hearing aid. Now an in-ear hearing bud. Now a cochlear implant ...

I realise, like it or not, we're already embracing the transhumanist project, we're just still calling it 'medicine.' We've inevitably blurred the line between 'therapy' and 'enhancement.'

Just before the 2008 Beijing Olympics, South African Paralympic runner Oscar Pistorius (who runs on curved carbon-fibre 'blades' and has been called 'the fastest man on no legs') entered the competition's qualifying rounds, pitting himself against able-bodied athletes. He didn't quite make the grade, but he wasn't far off. Hugh Herr, himself a double amputee and bionics researcher at MIT, says, 'It certainly will be possible in the future to build a limb that outperforms an intact limb, especially for running.' In fact, there is an ongoing argument among sports officials that Oscar's prosthetic feet give him an *unfair* advantage. It's conceivable that by the Olympics of 2012 or 2016, the 'handicapped' will be unbeatable. In laboratories around the world, research is leaping ahead, building limbs that are bonded to the body, controlled by the mind, stronger and lighter than the real thing, with artificial pressure-sensitive skin.

The Bionic Man isn't fictional any more. Considering the growth in cosmetic surgery it seems highly likely that in the not too distant future someone will elect to have a perfectly healthy limb deliberately removed and replaced with a 'better' artificial one. Unlike natural evolution, which requires centuries to fumble an improvement, our man-made technologies quickly keep on doubling in power and halving in price, ushering the prospect of the 'disabled' out-performing what we now consider the able-bodied. Will I be buying enhanced cochlear implants in a few years the way I bought a mobile phone in the nineties? Will I buy a new body part grown from my own stem cells to keep me young? How far can this go? How long can I live? How enhanced could I be?

Nick Bostrom suggests that the answers to these last two questions could respectively be 'a very long time' and 'as much as you like.'

Take, for instance, the idea of 'longevity escape velocity,' seen by many transhumanists as one of the more convincing arguments for the prospect of immortality. This is the proposition that instead of our life expectancies gaining a quarter of a year for every year that passes, we reach a point where medical advances raise our life expectancy by *over* a year for every year that goes by.

Bostrom paints a mind-blowing scenario. 'Once you have a sufficiently long life span then you would think that within that time there would be even more powerful technologies which can extend your life further. You can imagine that there will be ways of uploading into computers and then we'll have a space civilisation where you could live for billions or trillions of years. So if you could survive for the next several hundred years (or it might just be the next fifty, we don't know), at that point your life expectancy might become literally astronomical and, yeah, that would be a profound change to the human condition.'

Fly. Windshield. What does that kind of future *look like?*

'In my view it would be a distraction to focus on external things,' he says. 'It's better to think about what kind of mental state or mental activity would most correspond to what you would want to do if we had the ability to do whatever we wanted. So maybe try and find the highest reaches of human happiness and creativity and relationships – which will be refined even more in a posthuman future – but find the closest we can get today. What do you think the ideal is? Maybe the simple pleasures of family life, the ecstasies of artistic creation, romantic love, happy children eating an ice cream in a playground? I don't know. But whatever you think the ideal is, then one would think a posthuman condition would be able to realise that ideal and other ideals to a higher degree. That's the whole point of trying to get there – that it's something we think it's supremely worth getting to.'

I'm struck by how very *human* Nick Bostrom's motivations are – on an emotional level they are the things most people care about: relationships, love, great music. I try to imagine myself 'posthuman.' Might I look back on this book with quaint amusement? – the disjointed ramblings of an un-enhanced mind, stalked by mortality – caught in his own transience, trying hard to understand 'the big picture' but really one of Bostrom's ants, behind it all scared – literally – to death?

Short of an accident, I'm going to live longer than I thought. And I'll probably live healthier for a greater proportion of my life than my ancestors. I now believe that, at a minimum, stem cell therapies will become mainstream before I'm a pensioner – offering a very real possibility of repairing my ageing human body, potentially pushing me into my second century.

Just a couple of months after my visit to Oxford, scientists at Imperial College London announce that by stimulating the bone marrow of mice, they can release a flood of adult tissue-specific stem cells into the bloodstream. No invasive surgery, no cultivating cells outside the body, no baby parts. Dr Sarah Rankin, who led the research, likens the process to sending a fire engine to every street corner: 'Once they're in the blood they're "patrolling the streets." Just as a fire engine passing a house on fire would naturally stop, likewise your stem cells will go to an area of damage.' Rankin estimates human trials will begin 'in the next five to ten years.' Invasive surgery may become an increasingly rare occurrence.

Is it possible that in the future, if I need a hip replacement I'll grow my own? Researchers at the Columbia University Medical Center in New York have already demonstrated just such a technique in rabbits, using the animals' own stem cells to regrow joints. Elsewhere Dr Thomas Einhorn at Boston Medical Center has helped over fifty patients with degenerative hip disease by using their own stem cells injected into the hip, to help generate new bone.

That evening I visit the home of an old friend who lives in Oxford with her husband and their baby daughter. One-year-old Abigail Billen is quite possibly a member of the last generation who will substantially worry about disease and ageing (either that or they'll be ruled over by a race of genius-immortal mice).

Their world is going to be a very different place from the one we live in today. At the very least it will need a new moral framework to deal with the results of biotechnology, new ethics of enhancement, new definitions of the word 'therapeutic.' She will almost certainly view her mortality quite differently from the way her father sees his.

'After you're forty, I find, all you really think about is death,' my friend's husband says. 'Your perspective completely changes. You wonder, "Shall I bother to paint the bathroom door? – because I might be dead quite soon." Before I was forty I never factored that in. But after forty I worried about it. It's just a background hum all the time.'

He's not joking, so I'm glad he at least bothered to cook the rather excellent casserole we're tucking into. Surely here is someone who might actively embrace the transhumanist dream? What if he could be freed of that 'background hum'?

'Oh, I don't know,' he says. 'These people who want to live for ever are tremendous egotists. Life is like a sonnet. It's fourteen lines and you know how to operate within its perimeter. It's the rules, isn't it?'

But is it? Even now? For I myself am alive because the rules changed.

One Saturday at the start of the school summer holidays in 1981 my eldest brother was tasked with looking after me while our parents went shopping. He wasn't happy about this as he'd arranged to drive over to see his new girlfriend.

We never made it.

Ironically, if Ian hadn't spent so long making sure my seat belt was a good fit, we would have missed the car that went straight through the Stop sign and ploughed into our passenger side. The engine lifted off its mountings, twisted and landed on my legs, snapping both my femurs clean in half. My spectacles were found thirty feet down the road and the crash was heard a mile away. There was a lot of blood (mine).

My brother managed to crawl out across what was left of the bonnet but I was trapped in the wreckage. As I approached the crucial three-hour point of being pinned in, the paramedics decided the only way to save my life was to amputate both legs, which were on the brink of dying and taking me with them.

Then an ambulance man to whom I owe a great deal said, 'If we damage his legs trying one last thing he hasn't lost anything has he?' Dosed high on morphine I don't remember the fire engine and the ambulance attaching the ropes to either end of our fatally crumpled vehicle and literally pulling it apart.

I do remember being lifted out.

In the hospital, I received, I think, four pints of blood. Without it, the rescue effort would have been worthless. I didn't know it at the time, but I should have been thanking a nineteenth-century physician called James Blundell who performed the first successful human blood transfusion – and changed the rules about who lives and who dies.

I tell the story because I'm trying to imagine what would happen if any modern politician suggested we end the practice of blood transfusion. Medicine has been changing the rules ever since the first patient was saved, and we've welcomed it. Like it or not, we're not getting off this road, because we don't want to. Would any of us deny the newly windpiped Claudia Castillo the right to go dancing, or run in the park with her children?

It's quite likely that I'll live to hundred – and while that may not be long enough to get me into the running for becoming posthuman (although it might) it's longer than I was expecting and gives me a good frame of reference for my travels – what's going to happen in the next sixty years. It also means that if I have kids, I'll probably live long enough to see *their* children enter adulthood. And what my grandchildren might become is an interesting prospect.

Juan Enriquez, founding director of Harvard Business School's Life Sciences Project argues that our change as a species will involve the 'accumulation of small, useful improvements that eventually turn homo sapiens into a new hominid ... a new species, one with extraordinary capabilities, a homo evolutis.'

Every advance in therapy is also a step toward the posthuman condition. The transhumanists know this. The rules *are* changing. Medical science is sounding the death knell of the old game. Aubrey de Grey has said his work isn't about keeping people alive longer *per se* but 'to stop people getting sick.' Radically longer lives are, he suggests, really just a 'side effect' of good medicine. Such future medicine, I'm beginning to understand, is based on the ability to understand and manipulate our genes and the cells that carry them – the oft heralded but poorly understood 'biotech' revolution.

To understand the future – and get a grip on whether the transhumanists really are on to something – I need to understand biotech. It's both a wonderful and frightening world. The next stage of my journey will take me not only out of the country, but way out of my comfort zone.

CHAPTER 2

The most wondrous map

Life is rather like a tin of sardines – we're all of us looking for the key. ALAN BENNETT

On June 26, 2000, approximately 22.5 kilograms of carbon, 70 kilograms of water, 2.5 kilograms of nitrogen, 1.4 kilograms of calcium, and a half-kilogram mix of fifty-four other elements (including just over a tenth of a milligram of uranium and enough phosphorous to make twenty-five hundred cigar-lighting matches) walked into the East Room of the White House.

'Nearly two centuries ago, in this room, on this floor, Thomas Jefferson and a trusted aide spread out a magnificent map,' said then US President Bill Clinton. The map in question was the result of the first coast-to-coast overland exploration of the United States and the 'trusted aide' was Meriwether Lewis. But Clinton had come to talk about another map: 'Without a doubt the most important, most wondrous map ever produced by humankind.'

Seventeen years earlier, a man called George Church and a handful of his peers suggested making the map Clinton was referring to: a completed 'human genome' – a writing down of the genetic cookbook from which each human is uniquely baked.

It's a cookbook small enough to be stored in nearly every one of your cells but contains enough information to explain why one bag of mostly oxygen, hydrogen and carbon is Bill Clinton, while a similar quantity of exactly the same stuff is you.

The accolade for sequencing the human genome was grudgingly shared by Celera Genomics (co-founded by Craig Venter) and a publicly-funded international effort led by the National Human Genome Research Institute. Actually, Venter's team did it better, faster and over ten times cheaper. And they also tried to patent much of the knowledge.

Subsequent outcry from press, public and many scientists, along with some hastily prepared legislation, put the kibosh on such capitalistic ambitions. Celera's stock price plummeted. Venter got fired and he ruefully remarked later, 'My greatest success is that I managed to get hated by both worlds.' But he had helped give birth to a new era – one that he continues to play a pivotal role in shaping.

Even though you can't read the cookbook, your cells (and now human technology) can. So getting hold of a copy (otherwise known as 'having your genome sequenced') could soon play a part in saving your life. More than that, in fact. It may mean that you hardly have to visit the doctor, because you won't get ill in the first place.

Anyone on a quest to understand biotechnology at some point comes to the work of George Church – the man who helped kick off the human genome project all those years ago.

Church is a turtle-breeding, dyslexic polymath who is currently professor of genetics at Harvard Medical School. He is a man widely regarded as one of, if not *the* key architect of the unfolding biotechnology revolution – according to *Wired* magazine a man so smart that 'even the experts don't always get what he's talking about.' He hopes to usher in an era of 'personal genomics' – where medicine and health care are tailored to the individual from the day you're born. To that end he's set in motion the audacious 'Personal Genome Project' (PGP) – a study that will change medicine.

With some trepidation I phone the professor's number in Harvard and get straight through. 'Sure,' he says, in response to my request for an interview, but first he wants to test out my credentials. 'Give me an example of something funny. You're a professional, right?' I suggest there's a certain irony in research that implies sufferers of Parkinson's disease could be helped by taking Ecstasy (usually a drug associated with *getting* people moving, not stopping them) and he seems satisfied.

But before flying to Boston, allow me to digress, briefly, to the picturesque university town of Tübingen, located about twenty kilometres south of the German city of Stuttgart. My journey into the stuff of life starts, in many ways, there – for it was in Tübingen in 1879 that the most powerful alphabet known to man was discovered.

The man to write it down was one of the world's first biochemists, Albrecht Kossel. Except he didn't know he'd done it. He just knew he'd found four distinct substances within his colleague Johannes Friedrich Miescher's discovery of nuclein – a gloopy residue taken from the innermost part of human white blood cells. Miescher had extracted his nuclein from pus scraped off wounded soldiers' bandages. Kossel called the four substances adenine, thymine, guanine and cytosine, and sometime later got a Nobel prize, in part, for doing so.

One letter (A, T, G or C) is all you need to represent each of the substances Kossel discovered. 'A,' for instance, is shorthand for the arrangement of five atoms of carbon, five of hydrogen and five of nitrogen that make up a molecule of adenine. Along with T(hymine), G(uanine) and C(ytosine) you have an alphabet of just four symbols – and with that alphabet, amazingly, you can write out a representation of the basic instructions to make a human being, your 'genetic code.' It is these four substances that nature uses to store all her recipes. My friend Mitch tells me, 'I remember the four letters by reciting "All That God Created."'

It takes over three billion adenines, thymines, guanines and cytosines lined up in a row and in the right order to make the human recipebook as we know it – 'the human genome' – a row of chemical letters you'll find in nearly every one of your trillions of cells (the most notable exception being the red blood cells). Written out as A's, T's, C's and G's on paper, it'd take up roughly the equivalent of two hundred volumes the size of the 1,000-page Manhattan phone book, and take nine and half years to read out loud (assuming you read ten letters a second and never slept, ate or went to the toilet while you were doing it). Which in our computer age isn't actually that big a deal. It's roughly three gigabytes of data – just over twice the amount in an iTunes movie download of the genetic comedy *Twins* starring Danny DeVito and Arnold Schwarzenegger (whose scriptwriters, by the way, must have glanced over the science and said, 'well, we won't be needing that!').

Nature also provides a built-in backup of all that data. Adenine and thymine, and guanine and cytosine are like two couples who hang out but never swing. Adenine will only get into bed with thymine, and guanine only has eyes for cytosine. Within the confines of a cell, A's will snuggle up to T's and G's will soon form trysts with C's, one side reaching out with a 'hand' made of hydrogen to cop a feel of the other's nitrogen or oxygen (this, appropriately, is called 'hydrogen bonding'). The result? Two lines of data, one a 'negative image' of the other. (Molecular biologists call the original line of molecules the 'sense' strand, and the mirror image the 'antisense' strand.)

If you add two long poles of sugar and salt for each line of molecular lovers to anchor themselves to (and so stand firm in their embraces) and give the whole shebang a twist, you'll end up with something that looks like a spiralling ladder, its rungs made of adenine molecules holding on to thymine molecules and guanine molecules holding on to cytosine molecules, like billions of trapeze artists frozen in mid-clasp.

There you are, then. The famous 'double helix.' The most famous molecule in the world. A cell's individual hard drive. The code of life. A constantly referred-to book of recipes found in nearly every single cell that, amazingly, includes instructions to make everything *in* every cell. Your genome, encoded in the fêted and misunderstood 'Deoxyribonucleic acid' – the bane of unwitting criminals who leave it behind at crime scenes, the premise for *Jurassic Park*.

It is a dead collection of atoms, bereft *of* life, but the cookbook *for it*. Chemists call it 'chemically inert'. It is no more life than a recipe is food, than a biography is a person or a page of musical notation is *All You Need is Love*. But it is utterly universal, as far as life on Earth goes. While the data in your genome is different from that stored in the genome of, say, an African savannah elephant, both you and it share the use of this chemical alphabet. In fact, we not only share it with twelve thousand kilograms of unforgetting, floppy-eared *Loxodonta africana* but with every living thing on the planet – meaning the same four letters can be used to represent recipe books for a whole menagerie of life and disease.

This leads to a startling conclusion. If all life (plants, animals, bacteria, elephants, human beings) uses this coding mechanism, this means either whenever life gets going it somehow evolves *exactly the same coding mechanism* (coming up with the ATGC alphabet independently) or we're all descended from a common ancestor. The first cell. A third explanation (handy if you don't like the idea of admitting you're related to a fungus) is that God independently handed out the ATGC code to us all.

Wherever it came from, this is the code our cells read in much the same way components in your game console, deep down, read a code of electrical and magnetic charges represented by a two-symbol alphabet of 1's and 0's. When you think about it, that's a pretty arresting thought. We're used to putting code into a machine and getting out a game called '*Sonic the Hedgehog*.' But the ATGC code can give you *an actual hedgehog*.

Why those instructions are arranged the way they are and what happens to them inside the cell is something Professor Church has been working to understand for most of his life. In doing so, he hopes to revolutionise medicine.

I arrive in America the night before Labor Day (like William Shatner, a Canadian import) and, next day, with everyone on holiday explore the good city of Boston as the morning warms into a display of pure summery goodness. The only thing that's bothering me is why everyone I speak to asks me where in Australia I'm from. In the Public Gardens I come across a demonstration calling for health care reform. It's interesting to think that many of the medical conditions these demonstrators worry about could become a thing of the past if work like George Church's continues to bear fruit.

With the sun setting, I approach my hotel, passing the Broad Institute – a joint initiative by the mighty universities of Harvard and MIT. It is one of the many labs around the world spewing out documents full of A's, T's, G's and C's – labs that are writing down and cataloguing the DNA of pretty much anything they can get their hands on (including, in the Broad's case, that elephant I mentioned earlier). Currently, as a planet, we are collectively in the process of writing down the code of life in all its forms, including some that are extinct (their DNA being sifted out of hair, hooves and tusks held in museums).

The reasons for this gargantuan effort of genome librarianship are numerous but include, among other things, conservation. For instance, the University of Washington's Center for Conservation Biology have built 'DNA maps' that highlight genetic differences between elephants from, say, Zambia and Malawi, helping them pinpoint where a particular specimen has come from. Using this information, they can work out the origins of illegal ivory – and in 2009 were able to direct law enforcement to the borders of Tanzania and Mozambique as a result.

Elsewhere, the genetics research powerhouse The Wellcome Trust Sanger Institute in Cambridge, UK, is using genome sequencing to get new intelligence for another battle – that against cancer, taking the immortal words of China's most famous military strategist Sun Tzu to heart: 'know your enemy and know yourself.' Erin Pleasance and her team at the Sanger Institute examined entire genome sequences taken from two sets of cells from the same patient, one set ravaged by lung cancer (know your enemy) and the other as yet uninfected (know yourself). By comparing the two, they found that cells with the cancer had 22,190 DNA mutations (for example, a letter duplicated, deleted or replaced by the wrong one). That averages out at more than one mutation of your code for every packet of cigarettes. (These changes occur because there are over sixty chemicals in your smokes that bind to, and mutate, your DNA.) That's still an infinitesimal percentage of the genome, but just like missing the letter 'l' out of the word 'public' in a government white paper, the result can sometimes be a whole host of trouble.

Webb Miller and Stephan Schuster of Penn State University are using a similar approach to protect the Tasmanian devil from a mysterious and contagious tumour that is endangering the species. 'We are sequencing two specimens, one with the disease and another that seems immune, and hope to use the differences to guide a breeding program,' says Miller.

A final motivation for collecting all this genomic data is similar to the reason chefs buy lots of cookbooks – to add to their repertoire of dishes and get ideas for how to cook new ones.

The next day I head to the Department of Genetics at Harvard Medical School in the Fenway area of Boston. It's a slightly oppressive slab of glass and concrete, sited, appropriately, on Avenue Louis Pasteur – named after the iconic Frenchman who gave us pasteurised milk as well as the experiments that would prove to the world that diseases could be caused by 'germs' (critters like the bacteria and viruses that George Church's work has helped unmask and force to give up their genetic secrets).

As I approach I'm assaulted by a plethora of signs that announce, sombrely and rather patronisingly, 'Let's Be Clear. Harvard Longwood Campus is 100% Smoke-Free Both Inside And Outside.' While there's mild irony in a medical school having to ban smoking, what really makes me chuckle is that alongside these words is *the world's most pointless map*. On it, campus buildings are coloured dark blue, while the rest of the map is a lighter shade of the same colour. The two-part key has a dark blue square, labelled 'inside,' and a lighter one, labelled 'outside.' Handy for the world's smartest minds.

I meet Professor George Church inside. He is as welcoming a scientist as you could hope to come across – a man you'd be delighted to meet at a barbecue. When he talks, his tones are measured and gentle. That's not to say he won't challenge you or disagree about something, just that I get the impression that for George a disagreement is a chance to learn something, either about the subject under discussion or the person he's talking to. (During our initial phone call he'd almost warned me about this. 'You'll find I'm a little *too* curious.')

Physically he's a big chap. Six feet four inches tall and 245 pounds. That's slightly bigger than Muhammad Ali at his career heaviest, but George is almost the exact opposite. People often describe him as a teddy bear, and while that might seem overly cutesy, it's not far off the truth. He has a wide, open face, pale skin, a bushy uncle's beard and thoughtful eyes that seem to say, 'Well, isn't this world interesting?' He works for the enjoyment of discovering new things. 'I have very expensive hobbies,' he confesses. 'Synthetic biology and personal genomics.'

I can tell you a lot more about George you probably don't want to know. He has high cholesterol as a result of hyperlipidemia (for which he takes a drug called Lovastatin). Sometimes he suffers from an inflammation of the iris (a condition called, appropriately, iritis). He takes daily multivitamins as well as 100mg of 'Coenzyme Q10' – a supplement that's been linked to lowering blood pressure, helping DNA resist breaks and lengthening the life span of rats. He's right-handed. His blood type is O positive. He's also a narcoleptic (a disorder characterised by excessive daytime sleepiness).

I don't know all this because a) I am inordinately nosey; b) George is a vocal hypochondriac; or c) we had the world's most boring conversation ever, but because Church has happily published this data, along with his genome, online for the world to see as 'Participant #1' in what he describes as 'possibly the broadest, most invasive study ever in genetics' – the aforementioned Personal Genome Project – a study that points to the future of my healthcare more clearly than any other to date.

Over the coming years, the study will sequence a hundred thousand volunteers' genomes and try to correlate that data with their personal traits, their lifestyle and their medical history. It is the latest attempt to shed light on what is often oversimplified as the 'nature vs. nurture' debate. After all, your genome is one thing, but you are another – as identical twins amply demonstrate.

Identical twins start out at conception with exactly the same DNA, and the correlation between their genome sequences remains as close as a gnat's whisker for the rest of their lives. But we all know that even identical twins can turn out differently. One twin leads a healthy life, exercises, doesn't smoke, eats a balanced diet. The other has a twenty-a-day habit, thinks whisky really is the water of life and eats fatty foods. You don't need a degree in medicine to spot which twin has which lifestyle. Same starting point, same code – different lifestyle, different outcome.

However, it's rare for twins to act as differently as this, as two-thirds of the Bee Gees so amply confirmed, making it much harder to work out the subtle interplay between a Bee Gee's genome and his lifestyle. Is the ability to sing higher than a choirboy overdosing on helium a result of genetics, a diet rich in salami, long hours of practice in tight trousers, or a combination of all three?

More seriously, what's the web of interactions between our genome and our day-to-day lives that triggers disease? Why is it that your sister always gets colds but you rarely do? Did Uncle Bob get heart disease due to one burger too many? Was it always written in his code? Or is he the victim of some yet unfound link between eating mustard sandwiches and working in a laundry? Why do some of us develop an allergy to latex? Why can a bee sting send skateboarding legend Andy Kessler into fatal cardiac arrest, but not me – and why is it the more times you're stung, the more likely you are to develop a potentially fatal allergy? Or to put it another way, how *does* Keith Richards stay alive?

At one end of the scale there are some conditions that are 'in the genes.' This is genetic determinism that's as unforgiving as Richard Dawkins in a bible class. Huntington's disease (which famously killed folk legend Woody Guthrie in 1967) is a good example. In most cases, sufferers will lose their coordination, memory and ability to concentrate – declining slowly to death. Changing your diet or environment won't help you in the long run. Currently, it's incurable.

Then there are conditions that are still unavoidable, genetically speaking, but you *can* do something about, lifestyle-wise. One is Følling's disease, also known as 'phenylketonuria' – a disease which can amass too much of a substance called 'phenylalanine' in the body, with devastating effects.

'Phenylketonuria,' says George, proving he's a biologist by pronouncing the word without a flinch, 'is something every newborn in a hospital in the developed world gets tested for. If you've got it, you have to keep away from phenylalanine and aspartame.' That means steering clear of most artificial sweeteners, meat, poultry, fish or anything made with dairy products. If you don't, you'll get brain damage. 'It's hard, lifestyle-wise, but you can do it, and it changes the outcome from complete mental retardation to having normal brain development,' explains George.

And, as we all know, personal lifestyle choices (beyond those forced on us by conditions like Følling's disease) can also affect your health. Ninety per cent of lung cancers, for instance, are linked to smoking.

However, things are rarely this simple.

It's important (if you're working on curing disease) to know that the cookbook of your genome is split into twenty-three sets of paired volumes (one set from your mum, one from your dad) called chromosomes. Within these volumes (a full set of which is kept in most of your cells) are recipes for cell components that your biological machinery constantly reads and then uses to make whatever those cells need.

The content of both sets is roughly the same but in any pair, one volume (the 'mum volume' or the 'dad volume') will often have a 'preferred' recipe for a particular cellular component – this is the 'dominant' version and is considered the definitive article in cell land. If the copy in the other volume is 'recessive,' it is generally ignored. However, you can have two copies of the same 'dominant' recipe or two copies of the same 'recessive' recipe (if you got the same data from mum and dad), in which case either volume can be referred to with equanimity. These paired recipes are called 'alleles' (from the Greek *allelos*, meaning 'other'), and the mix of recessive and dominant alleles (recipes) throughout your genome will largely determine the mix of your parents that is you: how tall you are, the colour of your eyes, whether you can smell asparagus in your wee, whether you got mum's red hair, what diseases you're susceptible to and so on.

Each recipe is generally called a gene, although to be completely accurate a gene is *both* of the paired recipes ('gene' refers collectively to both 'mum-given allele' and 'dad-given allele') and it is the preferred allele that actually provides the instructions your cell uses. That recipe is usually for a protein – a protein being biologists' blanket term for a wide variety of cell parts.

Yes, on top of the lurid car-crash of syllables biologists so love, when it comes to naming things they, like some politicians, can sometimes use a simple word you *do* recognise to mean something *entirely different*. When biologists say 'protein' they're not referring to a nutrient found in meat, fish and eggs. For them a protein is a catch-all way to describe a whole catalogue of cell parts. Some proteins help us process food like the hormone insulin that plays a large part in turning sugars into cell fuel. Others manage chemical reactions like the enzyme telomerase that transhumanists get all excited about. (If you want a little more detail about why telomerase is causing such a fuss in transhumanist land, see the appendix at the end of this chapter – *Damned clever, those Russians*.)

A protein is the dish that gets cooked by following the recipe stored in an allele, chosen from a pair of recipe versions called a 'gene,' itself taken from the volume of a cookbook called a 'chromosome,' plucked from within the culinary *meisterwerk* that is your genome – a cookbook encoded in the four molecular letters of genetic code that we write out as A, T, G and C and stored on a hard disk called 'DNA.'

Having a particular arrangement of genes (or particular gene mutations) rarely *insists* you get a disease, it just changes the *chances* of you getting it – sometimes up, sometimes down. In most cases several genes have a role to play, all of them the genetic equivalent of a new card in a game of Texas Hold 'Em. Just as in poker, it's very rare that the cards you get will *absolutely* determine how the game will go; however, they nudge the odds with varying degrees of force in a particular direction. Your environment and lifestyle are cards in the game too. Vegetarians who don't get enough calcium increase their *chances* of osteoporosis. That said, those who avoid meat *tend* to get colon cancer less because they don't eat animal fats.

It's the same with your physical traits. There are very few one-to-one relationships between specific genes and traits (dimpled cheeks and your blood type are two of only a few notable exceptions). Your height, for instance, is dependent on the interplay of several genes, and (again) your diet.

Understanding the steps in the genome/lifestyle dance is, to a large part, tied up in understanding how many times the cell is referencing different recipes in the genome – 'gene expression' in the parlance of biologists. If it's cooking up four hundred copies of recipe A for every three copies of recipe B, this can tell you something. It might tell you what you've been infected by and how badly. (This is because cells are easily distracted by foreign DNA that's been injected into them by viruses, a process George will later summarise as, 'Oh, oh, look over there! Naked DNA! I love naked DNA!') Or it might let you know if you're in the danger zone to suffer from a particular condition.

For instance, 'over expression' of a class of recipe-genes called 'oncogenes' increases your chances of getting cancer. Oncogenes encourage cells to divide and proliferate—which is useful if you're a baby and want to get bigger, but less useful if you've got a tumour that wants to do the same.

If a gene is a recipe for a protein, then 'gene expression' is the number of helpings – and just as is the case with food, too much or too little of something can be bad for you. Nature and nurture. Genes and 'gene expression,' entwined like a couple doing the tango, an intricate dance of life where you cannot understand or appreciate one without the other. (Natalie Angier, in her balletic whirl around the fundamentals of the geek-sphere, *The Canon: the Beautiful Basics of Science*, neatly summarises it by saying 'you can't uncouple nature from nurture anymore than you can uncouple a rectangle's length from its breadth.')

Craig Venter told *Der Spiegel* magazine that so far the Human Genome Project has produced 'close to zero' medical benefits and that 'we have, in truth, learned nothing from the genome other than probabilities.' This is why the *Personal* Genome Project is so important, hoping to expose some of the more subtle steps in the interplay of our genetics and our lifestyles, to master some of that complexity.

George Church is less interested in the 'really, really sick' because it's not obvious to him that 'the mechanism by which you become really, really sick is relevant to most of the population.' The Personal Genome Project isn't really about saving the desperately ill (although it'll be useful in that battle too). It's about stopping most of us from ever getting that way.

It is arguably the most ambitious and most important medical study in history – one that could potentially inform the future treatment of thousands of conditions and reveal hitherto unknown knowledge about what makes the machine called *Homo sapiens* tick. The future of medicine, in terms of practical links between the *theory* of genetics and 'on the ground' medical practice, is being born here at Harvard.

If you take part in the study (and George wants you to), be warned: not only will you have key parts of your genome sequenced, you will also be measured in a myriad of ways (capturing your traits), asked a huge variety of deeply personal lifestyle questions and encouraged to divulge your entire medical record. You'll donate skin cells that George will reverse-engineer back to stem cells. And then you'll be asked to share all this with whoever wants it.

The reason is simple – the more people who have access to the full data, the greater chance there is of someone finding those links between our genetics and the lives we lead. The PGP will generate an enormous amount of information and George wants it interrogated as quickly as possible, by as many people as possible. The danger of one institution (even if it's Harvard) keeping the data to itself is, believes George, the danger that new knowledge will, at best, take a long time to emerge or, at worst, be missed altogether. 'You never know who's going to be the expert that figures something out,' he says. 'It could be some computer scientist in India, or some teacher in Bangladesh. I don't want to be the police officer that says: "Oh, you can't solve this problem. You don't have the right approval." The more people that are looking at it, the better.'

Maybe there's a link between Church's iritis, his turtle-breeding and a particular gene? One more genomed-sequenced turtle-breeding iritis sufferer in the PGP may be all that's needed to find out – and a budding biological Einstein, who could be toiling in obscurity, now has, via the Internet, the keys to a treasure trove of data. It is a level of data freedom that is almost unheard of in medical research, but as George says, 'There's a lot about the PGP which is far from business as usual.'

'One of the big things is finding people who have the same alleles,' he explains, referring to those paired 'dominant' or 'recessive' recipes. 'If you can find one case where someone is getting ill, but in the other they're not, then the difference is probably environmental. You can hunt down the environmental components and bingo! So, you change your lifestyle or your food and you get better.'

In principle, that isn't any different to stopping smoking to avoid getting vascular disease or lung cancer, but again it's the more subtle interplays which the PGP could expose that George is after, unearthing a whole bunch of personalised advice to keep us healthy. 'We can say "these are the recommendations for the environments that would be good *for you,* and here are drugs that *you* should and shouldn't take. Oh, and here are drugs you shouldn't take in combination."'

Talking of drugs, studies like the PGP could release an avalanche of lifesaving medicines that *already exist,* but aren't currently available. Right now the world is sitting on a wealth of useful drugs that never see the light of day. According to a 2001 study by the Tufts Center for the Study of Drug Development, just one in five thousand proposed medicines make it to being approved for human use by the US Federal Drug Administration. Even those that get as far as human trials have a less than one in four chance of making the final journey to being available for prescription.

The reason is that, although they may have benefits for many people, they can be dangerous, even fatal, to a small subset of others. Because we don't always know how to identify who will benefit and who will be at risk, drug approvals are necessarily cautious. One recent case in point was Vioxx, a popular brand of the painkilling drug Rofecoxib, often prescribed to patients with 'chronic pain' conditions such as arthritis. A study revealed that there was an increased risk (roughly doubling from a 0.75 per cent risk to 1.5 per cent) of heart problems if the drug was taken for eighteen months or more – so it was withdrawn under a heavy media spotlight.

The economic impact of this is that the expense of researching and seeking approval for drugs that never see the light of day has to be covered in the prices of those that *do* reach the market. This means the price tag to a drug company for each successful drug easily approaches a billion dollars (and can sometimes be twice that), which is one reason why drugs can be expensive. Personal genomics could change all that.

'Currently drugs are getting rejected because they fail in one per cent of the population,' says George. 'You can use genome data to figure out that one per cent and give the drug to the other ninety-nine per cent.' As well as saving millions of lives, this could potentially send the cost of drugs tumbling by allowing more drugs to hit the market.

Getting these benefits, however, requires accelerating the adoption of a key technology.

The personal genomics revolution depends on cheap, mass-market human genome sequencing. 'It's just really a matter of cost,' says Church. 'We should have been bringing the costs of sequencing genomes down to an affordable level, and that's what I've been working on for thirty-two years.'

This ambition is why, for instance, George made the designs of his genome sequencing machine – the Polonator G.007 – open source. The machine severely undercut its competitors on cost/performance, so he was giving away a commercial advantage. There's nothing to stop someone else copying and selling an exact clone, or using George's designs to inform their own. For Church, though, that's not the point. The quicker genome sequencing becomes commonplace, the quicker lives will be saved. This is why you'll find George advising so many competing firms, even founding them. Speed, and not money, is his motivator.

It's therefore heartening to find out that the cost of genome sequencing isn't so much falling as *plummeting*, showing the same sort of acceleration in price-to-performance we're now blasé about when it comes to computers. 'Craig Venter's first complete human genome sequence cost a hundred million dollars, layered on top of the three billion spent by the competing publicly funded effort,' says George.

Yet, as I write, in mid-2010, a firm called Knome (co-founder, George Church) will charge you just under a hundred thousand dollars to do the same. Nine teams worldwide are currently competing for the Archon X Prize in genomics which will award ten million dollars to the first team 'to sequence 100 human genomes within 10 days or less' at a cost of 'no more than $10,000 per genome.' (I don't have to tell you who's leading the team most people figure will win.)

But the ambitions of the prize may already seem out of date. In 2009, California-based Complete Genomics (yes, Church is again an adviser) announced that it could sequence a complete human genome for just five thousand dollars and thinks it's on track to do it for a thousand (although there's no commercial offering as yet). Already there's talk of the hundred-dollar genome. Oh wait, David Weitz of Harvard reckons he can do it for $30. If genome sequencing carries on progressing as it has, the ten-dollar genome sequence will soon be with us and one-dollar genome not long after. 'DNA sequencing costs were halving roughly once per eighteen months (like computer hardware) since the 1970s,' says Church. 'But over the past five years it's been halving every four months.' (This is just one of a number of startling examples of something called the 'Law of Accelerating Returns' that I will come across many times on my travels – and which has some interesting implications for us all.)

It's no exaggeration to say we're on the brink of a genome data explosion that will revolutionise medicine. The PGP is hugely ambitious in attempting to create a database of a hundred thousand, but imagine a world where we all have a document of our genomes in hand, or where the ability to sequence the genomes held in our individual cells and compare them (as the Sanger Institute's ground-breaking cancer study did) becomes common practice? This heralds diagnoses that go beyond 'Mr Stevenson, I'm sorry to tell you you have cancer' to 'Mr Stevenson, you'll be pleased to know that we've sequenced your cancer and therefore we have a personalised set of treatments for you.' Hopefully I should be able to avoid many conditions altogether by pre-empting them.

'Imagine the day when you and your doctor sit down to review a copy of your own personal genome,' says the blurb for the X Prize. 'This vital information about your biology will enable your physician to inform you of your disease susceptibilities, the best ways to keep yourself healthy and how to avoid or lessen the impact of future illness.'

In the same chapter of *The Art of War* that talked of the power of personal knowledge in battle, Sun Tzu also wrote, 'It is best to win without fighting.' John Wayne might have said 'Let's head them off at the pass.'

This is where medicine is going. And if you'll excuse the pun, George's church is a broad one. Because he's not stopping there. He doesn't just want to *read* the code of life. He wants to *write* it too. And that has both hopeful and terrifying consequences, as I'm about to find out.

APPENDIX

Damned clever, those Russians

'Every time a cell divides you lose little bits of DNA called telomeres at the ends of your chromosomes,' says my landlord Emmanuel, a man I've never met in person but who, by a stroke of inordinate transatlantic good luck, turns out to be an assistant professor at the very same Wistar Institute where Hayflick and Moorhead made their discovery of the Hayflick limit for cell division over fifty years ago.

'Telomerase addresses this problem by adding those short pieces of DNA back onto the end of the chromosomes,' he tells me. 'Think of telomeres like the plastic tips of shoelaces that protect the ends from fraying.' This would be all well and good if the amount of telomeres that were lost when a cell divided were completely replaced in the new cells that were created. But they're not. In fact, every time a cell divides, the two cells that result have slightly fewer telomeres than the single cell they came from. In a sudden insight while waiting for a train on the Moscow metro, biologist Alexey Olovnikov worked out why this was, linked telomere shortening to ageing and explained the cause of the Hayflick limit all at the same time. As James Bond might have said, 'Damned clever, those Russians.'

Olovnikov knew that as cells divide, their DNA must be copied into the two cells that are a result of that division. He began to imagine that the cellular machine responsible for copying the telomeres was a bit like a train passing over railway tracks. As the train rolls over the tracks, it makes a copy of the rails, but it needs to pass fully over a piece of track before it can copy it. At some point, though, the train hits the buffers at the end of the track. It can't pass over the last train-length section of rails underneath it, which means it can't copy them. The result is that the copy is one train-length shorter than the track it's just travelled on.

In cell land, telomerase is the engineer who comes and finishes the job off, marshalling telomeres together with a bunch of 'telomere-binding proteins' (called 'shelterin') to create the final cap that protects the ends of chromosomes. But our supply of telomerase runs down as we get older. The engineer hasn't got enough energy to do the job any more until eventually he retires and 'the telomeres get too short to protect the ends of the chromosomes,' says Emmanuel. At this point 'the cells turn themselves off and stop dividing to prevent genomic instability. The cell line begins to age and die.' It turns out that the number of cell divisions it takes for engineer-telomerase to finally stop turning up for work is between forty and sixty – the Hayflick limit.

The argument goes that maybe we can continually replace that telomerase, keep the engineer healthy and thereby lengthen life. Work by Yousin Suh and colleagues 'demonstrated that centenarians and their offspring maintain longer telomeres' compared with those of us who are shorter lived. Her team also concluded that people with genes that give them an extra-active telomerase-making system tend to enjoy a longer old age. People who maintain longer telomeres are 'spared age-related diseases such as cardiovascular disease and diabetes, which cause most deaths among elderly people.' (To be completely accurate there are many factors other than a lack of telomerase that contribute to telomere shortening, but that doesn't stop a boost in telomerase keeping our cells, whether healthy or cancerous, away from the Hayflick limit and 'senescence.')

Emmanuel tells me there is now a general consensus that there is a *direct link* between senescence and telomeres. In fact, in May 2010 scientists working for the BioTime corporation reported that 'by selecting for cells with sufficient levels of the immortalizing protein telomerase' they were able to 'reset the clock of ageing back to the embryonic state' and show that 'time's arrow of development, as well as ageing, could be reversed.' Like all companies with an eye on keeping their investors happy the BioTime announcement was dressed nicely. This is early-stage research with an interesting result, but a long way off any kind of application. Nonetheless, it does hint that the transhumanists aren't as bonkers as some people make them out to be.

CHAPTER 3

My lab should not be trusted

My momma always said, 'Life was like a box of chocolates ... You never know what you're gonna get.' FORREST GUMP

The biotech revolution goes way beyond the realm of human medicine – making it without doubt one of the most important stories of our age. Every living organism has a genome – as does every virus (the first genome to be sequenced was that of the snappily titled and bijou-genomed Bacteriophage MS2 virus, back in 1976). And it is these cookbooks that we're now learning how to edit: the 'genetic modification' that occupies so many headlines. The twin sibling of personal genomics is 'synthetic biology' – the tools, techniques and knowledge that today allow us to recode cells and, in the future, will enable us to fabricate entirely new ones to our own designs.

Such things, not unreasonably, give a lot of people pause for thought. Fiddling with life is something many of us believe should be reserved for more divine engineers than ourselves, the argument being that biology isn't a tool to be reprogrammed in the same way we load software into our iPhones. Others echo the view of biologist (and co-discoverer of the structure of DNA) James Watson who said, 'If we don't play God, who will?'

One man who, some critics argue, already thinks he *is* God is Craig Venter. As the CEO and chairman of Synthetic Genomics he has stated (with only a smidgen of irony) that he has the 'modest goal' of using synthetic biology to create new fuels and 'replace the whole petrochemical industry.' In May 2010, the newspapers announced he had 'created artificial life.' Venter himself was more measured. 'We've created the first synthetic cell,' he said. 'We definitely have not created life from scratch, because we used a recipient cell to boot up the synthetic chromosome.'

What Venter had done was to make a synthetic copy of the genome of a bacterium. The DNA sequence (which included several non-functional DNA 'watermarks' in order to distinguish it from the 'natural' version) was written in a computer and then manufactured out of synthetic DNA. This was then implanted into a bacterial cell, which divided billions of times, each daughter cell inheriting the synthetic DNA sequence. The newspaper flurry of 'Dr God' headlines both overstated and overshadowed an important and symbolic milestone in synthetic biology. George Church told me, 'I'm a big believer in lots of milestones and celebrations along the way. I think we should celebrate it as the first fully synthetic genome functioning in a cell, because that's what it is.' (In *Nature* magazine he noted that 'Printing out a copy of an ancient text isn't the same as understanding the language.')

It's with technology like this that Venter hopes to create genetically engineered bacteria that excrete diesel. He's not joking about the latter, and to prove it he did a deal with every environmentalist's favourite bad boy, Exxon Mobil, which is backing his idea with hundreds of millions of dollars. This kind of investment isn't an isolated event. San Francisco based LS9 (co-founder, one George M. Church) have modified the genetics of *E. coli* bacteria so that if you feed them sugar they produce fuels with 'properties that are essentially indistinguishable from those of gasoline, diesel, and jet fuel' – attracting the attentions (and money) of oil giant Chevron.

LS9's *E. coli* diesel project isn't science fiction. Far from it. You can actually visit LS9's pilot plant in San Francisco, pour a gallon of their bacterially produced diesel into your car and drive off. Construction of a new factory with 2.5 million gallons of annual capacity is underway. Want more? Venter has a life-form on the drawing board that makes fuel after eating carbon dioxide – a potential 'one stop' solution to both the climate and energy crises. When *Newsweek* asked Venter if his fuels would work in today's cars and energy systems, he replied, 'Basically everything we're making will work in the existing infrastructure.' He calls it 'fourth generation biofuels.' You call it petrol. Elsewhere, Joule Biotechnologies is *already* using genetically altered photosynthesizing bacteria to take in waste CO_2, mix it with sunlight and make diesel and ethanol. (Joule's scientific adviser? G. M. Church.)

Having just got my head around the idea that we can read our code (and the code of everything else), it now seems that there are plenty of people working on the idea of *rewriting* it – combining code from different organisms to make them do exciting and novel things. And with every organism on the planet being inexorably sequenced, there's one hell of a master cookbook of gene recipes to choose from and combine.

How about a bacterium that makes biodegradable plastic? Or one that pumps out components for anti-malarial drugs? Well, both already exist (both versions of *E. coli* again*). If you're diabetic, it's almost certain your insulin supply is now produced by *E. coli* bacteria whose genome has been tinkered with, and soon genetically modified plants will be producing it too. It takes just a short leap of the imagination (and, some argue, the technology) to design and breed a mosquito that immunises you against malaria rather than giving it to you (a bug you actually want to get bitten by) or to recode diseased cells to

* *E. coli*, it seems, is a family of bacteria that are a workhorse species in genetics. That's because they have a 'simple,' and therefore an easily fiddled-with, genetic makeup. Despite the bad reputation the bacteria get from a few black sheep in the family, it's generally a benign bug. You have a whole bunch in your gut right now generating Vitamin K2 (a vitamin also found in lots of those überhealthy dark green vegetables like spinach).

make them healthy (like rebooting or upgrading your computer). Venter's 2010 synthetic cell announcement foreshadows a world where we start to have the same control over biology that we have over software.

But as Franklin D. Roosevelt and Spiderman both knew, with great power comes great responsibility.

Consider Huntington's disease. If we could recode carriers' DNA to save them from its (currently inevitable) consequences, should we? I know if I had the faulty gene in question, I'd take a DNA upgrade – whatever the cost. And sufferers of Følling's disease would no doubt relish the chance of having some cheese on their crackers without, literally, losing their minds.

If we can recode people to avoid disease, we can probably recode them in other ways. There are genes that have been linked to intelligence, athletic prowess, ability to concentrate, aggression, even propensity to vote. It's wise to point out that many of these links are often more media-made than scientifically proven (the 'voting gene' being a key example), but there is a future coming where we could have the tools to nudge someone's abilities in one direction or another by fiddling with their code. It's a tempting treasure chest – or Pandora's box – of genetic riches which opens up what ethicists might call 'a huge can of (genetically modified) worms', amply demonstrated by the debate over what is known as 'gene therapy'.

To date, studies in gene therapy have tended to revolve around using viruses to deliver new or 'good' code into our cells. After all, viruses work by importing their own DNA into cells, so why not engineer them to carry helpful genes rather than harmful ones? Well, as *Bad Science* author Ben Goldacre might say, 'I think you'll find it's a bit more complicated than that'. The actions of viruses aren't the easiest things to control or predict – as the tragic case of Jesse Gelsinger demonstrates.

Jesse had mild ornithine transcarbamylase deficiency, a liver condition which causes the body to stockpile ammonia and can easily lead to brain disease. Newborns with the condition generally slip into a coma within three days, suffer brain damage and die in infancy. Eighteen-year-old Jesse was 'lucky' (in a relative sense) that his mild form could be controlled by diet. He was a willing guinea pig, understanding that the injection into his liver of an army of genetically engineered 'adenovirus' wouldn't necessarily cure *his* condition, but could potentially help in the search for a genetic therapy that might one day be used to treat newborns and other sufferers. 'What's the worst that can happen to me?' he said to a friend before heading to the hospital to take part in the trial. 'I die, and it's for the babies.' Within four days his 'worst that can happen' happened. His immune system took the virus to be hostile and went into overdrive, spiralling fatally out of control and resulting in multiple organ failure. Jesse is now something of a macabre *cause célèbre* for those who question the wisdom of attempts to recode our cells.

And you can see their point. Other dangers associated with gene therapy include the risk that the code inserts itself into the genome in the wrong place and triggers cancer; that a friendly virus doesn't know when to quit and does 'too much of a good thing'; or that a newly arrived gene gets 'over expressed' with harmful results.

It is over ten years since Jesse Gelsinger's death and gene therapy is still in the lab. Advocates point to successes like treatment for Leber's congenital amaurosis (LCA). Sufferers (who number about three thousand in the US) lack a critical enzyme in the cells of the retina. As the years pass, their world gets darker and darker. In 2009, using an adeno-*associated* virus as a carrier (one that has much milder immune response than the straight-up adenovirus that killed Jesse), the same university that hosted the Gelsinger tragedy injected a gene that codes for the missing enzyme into nine-year-old Corey Hass's retinal cells. 'Corey doesn't use his white cane anymore,' says his father, Ethan. 'When he plays with his friends out back and it starts to get dark, he can still play out there and see. It's pretty amazing. They've changed my son's life forever.' There again, Jesse Gelsinger's father could say the same.

It's important to realise that Corey's condition involves a *single* faulty gene (related to a *single* enzyme), that physicians know exactly where to deliver the virus (the area affected is a tiny section of the eye) and that cells affected by LCA, while not doing their job, are rare in staying alive waiting to be fixed – hanging around like lumberjacks who've lost their axes. Viral approaches are still a long way from working to cure multiple-gene conditions that affect larger areas of the body. In short, viral gene therapy has had its 'difficult second album' and is still waiting to see if it has a career.

In a paper published by George Church's lab shortly before our meeting, the situation was neatly summarised: 'Our ability to sequence genomes,' write the authors 'has greatly outpaced our ability to *modify* genomes.'

But all that could be about to change.

'Delivering by virus is an *easy* approach, but it's probably not the *best* approach,' declares Church. 'The ideal approach would be to take the cells out, reprogram them, do some quality control, make sure you've changed the right genes and put those cells back into the body.'

'Is that possible?' I ask.

'It's technically feasible, but it's not super practical.' Church pauses. 'Yet.'

Professor Church's proposed 'quality control' mechanism is the grandaddy of methods for sorting the wheat from the chaff.

Darwin's sieve. Evolution.

It won't rival Dan Brown for readership but 'Programming Cells by Multiplex Genome Engineering and Accelerated Evolution,' a paper co-authored by George Church in 2009, could turn out to be one of the most far-reaching documents ever written.

In it the authors describe how, rather than trying to change a cell's genome one gene at a time, they introduced *multiple* genetic variations (e.g., an insertion, substitution or deletion of a chemical letter) into the cell's genome. Each variation was designed to nudge the cell's behaviour in a desired direction (in the first instance the team wanted to see if they could increase production of lycopene, an antioxidant found most commonly in tomatoes). Over several generations of cells, a whopping twenty-four DNA modifications were introduced into the genome of *E. coli* cells.

Ever the engineer, George and his team built a machine to automate this entire process 'for *large scale* programming and evolution of cells.' Due to the method used to modify the DNA, different cells showed different variations to their genetic makeup, and each day the machine generated – wait for it – 4.3 *billion* variations of the *E. coli* genome. Within just three days of frenzied cellular reconfiguration, a small number of the nearly fifteen billion variants created were delivering the goods – a five hundred per cent increase in lycopene production. Cultivate *these* cells a few billion times and 'Job Done.'

It's not so much 'survival of the fittest' as 'survival of the cell that does more of what you want.' LS9, George's biofuel company that wants to engineer *E. coli* to make gasoline, is now using the technique to improve the efficiency of their process. 'What once took months now takes days,' says Stephen del Cardayré, vice president of research and development. A commercial version of the machine is in the offing. *Wired* magazine described it as a technique that 'could make it as easy to rewrite a genome as to read it.'

Now we're cooking.

If you're treating a disease – say, of the liver – you could use a technique like this to quickly generate an army of 'fixed' liver cells. 'So it'd go something like this,' says George. 'You irradiate some chunk of your body, killing off diseased cells. Then you inject the engineered donor cells and they go straight to that depleted area.' No DNA tinkering inside the body and no virus taxis. Instead, a batch of 'tested for health' cells ready to put you right.

'The irradiation bit doesn't sound fun,' I say. After all, irradiation is literally 'exposing something to radiation,' radiation that in this case is designed to kill tissue. It's something of a blunt instrument, too. 'It's not like you can pick off just the diseased cells, is it?' I ask. 'You're frying some good stuff too?'

George laughs. 'Yeah. Irradiating your body is not good for it. But you can imagine that with a little bit more synthetic biology, you could get these new cells to home in on where they're needed, and first *kill* whoever is there, but only kill them on a *need to kill* basis, right? You wouldn't need the irradiation.'

It's a staggering proposition, and George says it so calmly that you'd be forgiven for not grasping the importance of what he's saying. Like many engineers, George talks revolution as if he were fixing a plug, and I'm reminded of something I'd heard him say in a video of the 2009 Edge Foundation sponsored masterclass, 'A Short Course in Synthetic Genomics': 'We can program these cells as if they were an extension of the computer.'

'The only thing that puts this kind of medicine far away is really *will*,' says George, dropping for a moment his genial, relaxed, rationalist demeanour, showing a heartfelt emotion. 'The question is, how motivated are we?'

It's a hell of a question, because I suspect the answer is 'a lot' and 'no way, José' simultaneously. Because we're not just talking about curing disease. Sure, I would welcome a cure for Huntington's disease, and giving sight to the formerly blind seems only fair for those who want it. But I'm less sure about built-in night vision for soldiers. Gasoline without digging up fossil fuels? Bring it on! Then again, surely there's a whole new class of biological weapons in the chest too?

Suddenly my mind conjures up a strange mix of *Dr Strangelove* and *28 Days Later*. George is in the middle of the wasteland, decimated by rampant genetically engineered bugs that have destroyed the biosphere, leaving only him and the man who does movie voice-overs alive. 'If only he'd known . . .' he says, as the sun sets on a dead world. I'm also reminded of my chat with Nick Bostrom in Oxford. 'Maybe you're a happy little ant, building a good ant hill. But maybe you're contributing to Hitler's war machine?' I say to George, 'It seems to me that for every up side there's an equal risk. There's some serious bioterror issues here, aren't there?'

'Oh yeah,' he says. 'This is worse than chemical and nuclear weapons. Those weapons become less and less hazardous as they spread over larger and larger areas. But life? Well, a life-engineered weapon could become *more* hazardous as it spreads. You engineer the right virus, and infection can spread *exponentially*.'

This is seriously dark. I've got a knot forming in my stomach. 'Surely,' I hazard, 'it only takes one bad bug to get out, and we could all suddenly be victims to a new disease?'

'Yes, that's a totally fair scenario,' says George as calmly as if he's just asked me if I want milk in my tea. I'm admiring his candour on the dangers of his work, but the knot is tightening, twisting.

'*Doesn't that keep you awake at night?*' I ask.

He deadpans. 'I'm a narcoleptic. Nothing keeps me awake at night.'

It's a nice moment of levity, which fades quickly.

'We will take some risks,' says George. 'Maybe even some existential risks.'

'Existential risk,' it pays to remember, is boffin-speak for 'The End of the World As We Know It.' Bostrom explains it as risk 'where an adverse outcome would either annihilate Earth-originating intelligent life or permanently and drastically curtail its potential.' In summary: game over.

In the cheering paper 'Existential Risks: Analyzing Human Extinction Scenarios and Related Hazards,' written while he was at Yale, Bostrom refers to some Australian gene-fiddling undertaken in the search to create a new contraceptive. The result was a virus that killed every last one of those it infected and, worryingly, was highly resistant to vaccination.

It's tempting to claim that the search for a new Australian contraceptive was part of a larger program to engineer stay-cool beer and better cricketers, but it was in fact part of an effort to manufacture a pest control agent for rabbits. In the study, an extra gene was added into the mousepox virus. Despite its 'pox' label, the untinkered-with version of the virus is associated with only mild symptoms. But with a single new gene added, it became a laboratory-wide rodent genocide. Within nine days, every mouse who'd been infected was dead. In the exact opposite of what happened to Jesse Gelsinger (whose immune system went into overdrive) the part of the mice's immune system that combated viral infection simply shut down, leaving them defenceless.

In an interview for *New Scientist*, Ron Jackson, part of the team who ran the study, said that if 'some idiot' put the equivalent human gene into human smallpox 'they'd increase the lethality quite dramatically.' He continued, 'I wouldn't be the one who'd want to do that experiment.'

What was doubly worrying about this experiment was that its results went against a general trend. It's pretty much the case that if you're altering the DNA of viruses and cells in order to make them do new things, you generally make them less good at the things they used to do. As George points out, 'If I create an *E. coli* bug that makes gasoline, well gasoline is not good for most bugs, right? You can get them to survive but they're spending a huge fraction of their metabolism making gasoline rather than making baby *E. coli*. Compared to other *E. coli*, they're basically wimps.'

But with George's new techniques for generating billions of genetic variants, the possibility of finding a super-dangerous and robust bug rises dramatically. (By the same token, of course, the search for the super-safe bug is also made much easier.)

A budding bioterrorist (or just naïve biohobbyist) needn't go to the bother of engineering a new bug either. You can buy a secondhand DNA synthesis machine from eBay, while the National Center for Biotechnology Information (NCBI), just outside Washington, D.C., has the A's, T's, C's and G's for the 1918 influenza virus (body count: between 20 and 50 million) written out and freely available online. (It's worth noting that this code has been used to resurrect the previously extinct virus for research purposes, and it's now alive and well in laboratories in the United States and Canada.) In fact, the NCBI has hundreds of flu genomes available to download – another example of the world's current genome sequencing frenzy. Since 1982, genome data held in the NCBI's 'GenBank' has undergone exponential growth, doubling every eighteen months.

At a 2008 conference arranged by Nick Bostrom, with the child-friendly title 'Global Catastrophic Risks,' Dr Ali Nouri from Princeton University's Program on Science and Global Security summarised it nicely: 'What we worry about is that these tools will fall into the wrong hands and in some sense this is really an unprecedented threat – when a few people or even technically competent individuals gain access to tools that can produce so much damage.' In the book published as a result of the conference, Bostrom and Milan Ćirković wrote, 'The technological barriers to the production of superbugs are being steadily lowered even as the biotechnological know-how and equipment diffuse ever more widely.'

The 2001 anthrax attacks in America or the 1995 sarin nerve gas attacks on the Tokyo subway act as powerful markers in the public consciousness when it comes to biosecurity. As the technology to manufacture pathogens gets cheaper, it's a natural worry that either 'something will get out by accident' or that an ideology that sees mass murder as a legitimate action will gain the skills and technology to deliberately unleash a 'terrorist pandemic.'

Suddenly my future could be a lot shorter. One bad bug ...

'So what do we do?' I ask George Church.

What he says next surprises and heartens me. 'Well, we shouldn't depend on the goodwill of the scientists.'

Good start.

'We should have *active* surveillance.'

I lean forward. 'You want to be watched?'

'My lab should not be trusted,' he says.

'You're telling me you're untrustworthy?'

He smiles. 'No, I think we're *very* trustworthy, but there's no reason to believe me, right? So, it's just easier to put us under surveillance.'

'And you'd welcome that?'

'Yeah, we're inviting that. And on top of surveillance, the entire biotech industry should be licensed too. At the moment, we're at a dangerous situation because this stuff is easy to get hold of – and that has to change.'

Well, I'm glad he sees it. George certainly isn't falling into the category of the scientist too blinded by the pursuit of knowledge to see a bigger ethical picture.

And, now I think about it, he's not the only one. Nick Bostrom not only arranges conferences to look at this kind of thing, but his first introduction was to invite me to a seminar on biotechnology ethics. Elsewhere, Stanford University's Drew Endy, another high-standing synthetic biologist, is well respected for bringing ethical considerations to the fore. 'There is a lot at stake,' he says in an interview for the *Technology Review*. 'I really want to see how far this field can go,' but admits the answer will be 'not very far' if the industry doesn't grapple with the bio-error/bioterror issue comprehensively.

Before coming to Harvard I'd squeezed in a short trip to the University of Southern Denmark to meet Mark Bedau, not only a founder of synthetic biology pioneers ProtoLife*, but also a philosopher and an advocate for 'scientific social responsibility.' 'There are dangers, there are risks and you have to face them squarely,' he said. 'But the same is true with any other kind of powerful technology. This is why, from the very beginning, scientists in this field have taken the initiative to try and think through the implications, to think about how to do this in a way that is responsible. It's the scientists who are taking the lead.'

You'll find George arguing in print for 'a code of ethics and standards' and a 'list of precautions only limited by our creativity' in an article written for *Nature* four years before our meeting. 'Learning from gene therapy, we should imagine worst-case scenarios and protect against them,' he writes – a reference to the Jesse Gelsinger tragedy. 'This requires not just laws but monitoring

*ProtoLife's approach to synthetic life is different to the Harvard model. While George is making artificial cell components, he's using nature's template. ProtoLife comes at the problem from a 'first principles' standpoint, which says something like, 'Okay, we're going to create life, it has a container, it has some code, it has a metabolism, but it doesn't necessarily have to be based on DNA, or proteins, or the rest.'

compliance,' he continues – something he believes can be largely automated. After all, you cannot make synthetic DNA without machines and machines could have surveillance and safeguards built in. By 2007, George was taking part in a joint study between the J. Craig Venter Institute, MIT and The Centre for Strategic and International Studies along with a 'Who's Who' of synthetic biology. The study outlined seventeen options for governance.

'Can it be done?' I ask.

'Sure. Right now, there are bottlenecks which would be easy to license. So phosphoramidites are what you use to make synthetic DNA, right? You don't use phosphoramidites to make toothpaste, or lubricants, or anything like that. So if you license the few companies in the world that are making phosphoramidites then you're done, right? You make it hard to get a license and then you've a legal framework you can work with. Anybody making DNA using unlicensed phosphoramidites, or anybody who makes phosphoramidites without a license is in trouble. Another area is licensing the machines that make DNA. It's not easy to make them. If you tried to buy the parts, that would set up red flags all by itself.'

'Like if I tried to buy some uranium?'

'Exactly. You want to make it as hard as possible to do this on your own. And we already have frameworks for this kind of thing – for making pharmaceuticals, for instance. I think there's general consensus that this can be done, and fairly inexpensively.' He pauses. 'I realise this may not be an exciting answer . . .'

'But it's incredibly important,' I say, relaxing a little.

'It's an example of pragmatic optimism,' continues George. 'Scientists and companies should be willing to accept licensing. And if you license the whole ecosystem from the chemicals, the instruments, the know-how, the genes . . .'

As I write, a fairly bewildering array of consortia both public and private are in various stages of gestation, signalling the birth of the licensing regime George advocates. The US National Science Advisory Board for Biosecurity is fostering international dialogue of biosafety issues, as are both the World Health Organisation and the UN (the latter considering how the existing Biological Weapons Convention can be enhanced to encompass synthetic biology). The commercial syn-bio industry too has several overlapping initiatives – the International Association Synthetic Biology, the International Gene Synthesis Consortium and the Consortium for Polynucleotide Synthesis. In particular, members of these consortia are beginning to implement the scanning of DNA orders – making sure no one is ordering anything that looks like a pathogen. It's heartening to note that in covering Bostrom's Global Stratospheric Risks conference, *New Scientist* chose to write an article entitled 'The End of the World is Not Nigh.'

It strikes me that a race is on – between the science and our ability to implement a regime that will help us reap the benefits without unleashing a tirade of bio-horrors. Give us the medicine. Save us from the weapons. But if you can't save us from the weapons, are we sure the medicine is a price worth paying? Sadly there is no equivalent of 'multiplex genome engineering and accelerated evolution' for policy makers – giving us four billion biosecurity frameworks to choose from by tomorrow. As ever, the science is moving faster than we are.

So, we'll have an accident. Synthetic biology will have its Three Mile Island, its Windscale, its Chernobyl. And somewhere, some ideology-driven buffoon will succeed in a deadly bio-attack giving us synthetic biology's 9/11. I suspect we'll have done most of the things we can think of to prevent it, but it'll happen anyway. We'll license the phosphoramidites. We'll put chips inside DNA machines that send emails to the FBI if they're asked to make anything dodgy. We'll update our weapons treaties. But somewhere, something will go wrong.

It is now largely accepted that an overlooked piece of drain maintenance at the Pirbright animal disease research facility in Surrey, England, allowed an incidence of a 1967 virus no longer in circulation back into the UK countryside in the summer of 2007. The resulting foot-and-mouth epidemic dealt a sucker punch to the British cattle farming industry. Our landscape seemed littered with pyres of beasts burned to contain the spread of the disease.

One. Bad. Drain.

As with the foot-and-mouth epidemic in the UK, we'll most likely contain future pathogen leaks to one degree or another. And then we'll learn how to make sure it's less likely to happen again, responding better and quicker, not least because we'll be able to sequence our enemy, find its weak spot and release countermeasures (chosen perhaps from a palette of fifteen billion pathogen-bashers made within days). Disease screening is already a reality. The Institute of Bioengineering and Nanotechnology in Singapore have developed technology for a screening device that can identify multiple diseases. Used at airports, it could halt the spread of an epidemic in its tracks.

It seems likely, too, that there will be a revolution in protective medicine. The 2002 figures from the World Health Organisation show that cardiovascular, infectious or parasitic diseases and cancer account for sixty-five per cent of deaths each year. While it is not *impossible* to imagine an engineered pandemic virus suddenly taking out an equivalent thirty-five million annually, it is very hard to imagine this doing it year in, year out without some kind of fight-back from our medicine.

A good example of this fight-back capability can be seen in our response to the threat from the H5N1 'bird flu' virus. (Bird flu is one of the 'global catastrophic risks' that Bostrom's conference decided was more likely than others to put humanity in a tailspin.) Particularly worrying is the virus's ability to mutate (influenza viruses love to change the rules of engagement). And just like an engineered pathogen, a contagious version of bird flu would benefit from the increasing density and mobility of our populations, turning an outbreak into an epidemic at lightning speed.

But now we have the ability to sequence and start to understand how the various viruses in the bird flu family work, allowing governments to stockpile the most appropriate antiviral drugs. This also gives us a far greater chance of developing preventive measures. We can 'know our enemy' far quicker than we did in 1918. As of September 2009, the World Health Organisation reported that there are no less than a hundred and ten potential vaccines in development to battle a bird flu pandemic. During the 1918 flu pandemic there were none.

So while we might have raised the chance of engineering something nasty, we've more than compensated by developing techniques that could combat it. The same techniques will also help to see off things that *already* kill us today.

By the time I leave George Church, my mind is buzzing. *Everything* is changing. Or as he puts it:

> We will learn so much more about ourselves and how we interact with our environment and our fellow humans. … I am optimistic that we will not be dehumanised but rehumanised, relieved of a few more ailments, better able to contemplate our place in the universe and transcend our brutal past.

This isn't like learning how to build a better internal combustion engine, or the invention of the TV. This is engineering *life itself*. It's tremendously exciting, offering hope on the one hand and potential terror on the other – just like a great movie. But how will it play out? Are we heading for a disaster, or a disease-free utopia? Probably neither. History is rarely like Hollywood. (If it were, we'd all be a lot thinner and less intelligent.)

Our biology is on the brink of moving at the speed of change we're used to seeing in computer hardware. I can almost feel the world shifting under my feet and, as I head out into the early dusk, something George said to me toward the end of our conversation begins to echo round my head.

'Evolution is far from over,' he said. 'It's accelerating, and something is going to replace *Homo sapiens*. Is it going to be wetware, or is it going to be silicon? Is it going to be genetically engineered, or is it going to be electrically engineered? Or, is it going to be some hybrid?'

There was a pause.

'It's just really a matter of when.'

PART 2

MACHINE

CHAPTER 4

Can you turn all the buttons on?

Man has made many machines, complex and cunning, but which of them indeed rivals the workings of his heart? PABLO CASALS

'Press the red button.' Leonardo looks down. It's the sort of big red button that wants to be pushed – the sort of button that in old sci-fi movies launches devastating barrages of weaponry. It's not just a button. It's a statement.

Gingerly he places his palm on top of it, blinking. He presses it. It lights up and (because this is modern-day Boston and not a scene from *Flash Gordon*) nobody gets vaporised. Satisfied, he stares wide-eyed at his teacher, Andrea, awaiting instruction. Leo doesn't know it, but he is about to prove he's a genius.

'Press the green button,' she says. He obliges.

'Now all the buttons are on,' she explains. Leo nods. His moment of genius is approaching.

Andrea turns both buttons off, and Leo blinks twice in clear surprise. *She just asked me to turn those buttons on!* As he glances down to take in this wilful act of button extinguishing, she is putting something else in front of him.

'This is the blue button,' she says. Leo regards the newcomer. He moves his head from side to side. *Yes? What now?*

'Have you learned to turn all the buttons on?'

Ah! That game again. He nods confidently, reaches out to the blue button and, after a slight pause, pushes it down. She looks on impassively. Leo has a think.

And then it happens.

Tentatively he presses the red button. Then, finally he reaches out and presses the green button. He looks up with big adorable eyes, searching for approval. All the buttons are on.

'Very good,' she says.

Leo has just proved he's a genius.

Now, as great achievements go, you may think this is hardly up there with Einstein's Theory of Relativity, the Clavinet riff in *Superstition* or indeed that masterful goal by Pelé in the 1958 World Cup Final against Sweden, but what Leo has just done has some startling, possibly world-changing implications. Leo, you see, is a robot – and he's just done something robots aren't supposed to be able to do. *Think*. Or at least convince you that they're thinking.

In a 2008 article for *Scientific American*, Hans Moravec, research professor at the Robotics Institute of Carnegie Mellon University reminded us that robotics had failed to live up to the predictions of the 1950s: 'Within a decade or two, computer scientists believed robots would be cleaning our floors, mowing our lawns and, in general, eliminating the drudgery from our lives.'

It's not the mechanics of making a body that's causing the problems, argues Moravec. While there have no doubt been many difficult challenges in replicating or making equivalents of human and animal body parts, these are largely being met, driven partially by the development of new materials and ever faster and smaller electronics. The Shadow Robot Company of London manufactures hands that demonstrate the same dexterity and range of motion as their human counterparts, with the strength to firmly grasp power tools and the delicacy to hold an egg. In 2005, Boston Dynamics revealed 'BigDog' – a four-legged robotic packhorse that can carry 340 pounds and traverse tricky terrain with apparent ease. Robotic vision systems can now capture images in resolutions that rival the human eye. No, it's not the mechanics of the robot body that's the sticking point for the autonomous intelligent robot, it's giving it a mind.

Which is why what Leo has just done is so important.

According to his creators at MIT, Leo has just 'generalised a task goal to a new configuration.' That's geek-speak for 'working out how to do something from having done something similar, but not identical, before' (a bit like being able to play Ping-Pong because you've played tennis). It's a skill we learn as children and is essential for navigating the world around us. Leo appeared to demonstrate that he could understand that 'all' meant both the buttons in the first test, but *all three* buttons when the blue button was added. 'All' is an abstract concept, which changes depending on circumstance. To you and me, the ability to 'generalise' the concept of 'all' seems easy – although this kind of abstract thinking is hard won in childhood. Robots, like small children, find general or abstract concepts difficult. But if Leo is genuinely thinking, does this mean the intelligent robots are finally coming?

Leo doesn't look much like your average robot. If Yoda had slept with a German shepherd (the dog breed, not an *actual* German shepherd, you understand), the result would probably resemble this three-foot-tall ball of furry, fat-bellied cuteness. He has wide, expressive eyes, a small black nose and huge shaggy ears. He looks like a fantasy creature from a Spielberg movie, which isn't that surprising given that he was originally built by Stan Winston Studio, a creature workshop for films such as Spielberg's *Jurassic Park* and, aptly, the Kubrick/Spielberg movie *Artificial Intelligence: AI*.

The folks at Stan Winston Studio know how to make mechanical things look lifelike. But the animatronics in movies are human controlled. As complex as they are, as impressive and lifelike as they seem, they are superpuppets that 'die' the moment their masters break for a biscuit. Leo is different. As soon as he arrived at the Personal Robots Group in 2002, his animatronics were stripped out and an autonomous robot was put in their place. Nobody was controlling Leonardo in the button exercise. He was acting alone. Leo, like you and me, is his own puppeteer.

As soon as George Church invited me to Boston, I knew I wanted to see Leo's 'mother' – MIT roboticist Professor Cynthia Breazeal who, aged ten, saw *Star Wars* and dreamed that one day she might build something like C-3PO. Now, most ten-year-olds watching *Star Wars* want to be Luke Skywalker, Han Solo or Princess Leia. Those who instead see the hardware and think, 'I'd like to make one of those,' and forgo the lightsaber for the soldering iron, are just the sort of kids who end up at MIT.

Professor Breazeal is trying to build an autonomous, social robot – a robot we'd be happy to treat as a living (if not breathing) entity and not just an arrangement of servos, computers and sensors. She wants to create a personality we can interact and work with, even be friends with – to all intents and purposes, an intelligent, social, multipurpose machine. This scares the hell out of a lot of people. As Breazeal's former mentor, the restive, maverick genius Rodney Brooks, opens his book *Robot: The Future of Flesh and Machines*: 'As these robots get smarter, some people worry about what will happen when they get really smart. Will they decide that we humans are useless and stupid and take over the world from us?'

We're weaned on a diet of Hollywood movies that generally cast future robots as perilous weapons of war and aggression (*The Terminator*) or scheming murderers (like the iconic HAL in *2001: A Space Odyssey*), with only the occasional *Wall-E* as light relief. It's hardly surprising we fear being dominated by our mechanical progeny when our dreams are so dark. Noel Sharkey, professor of Artificial Intelligence and Robotics at the University of Sheffield, told *New Scientist*, 'Isaac Asimov said that when he started writing about robots, the idea that robots were going to take over the world was the only story in town. Nobody wants to hear otherwise. I used to find when newspaper reporters called me and I said I didn't believe [that] AI or robots would take over the world, they would say "thank you very much," hang up and never report my comments.' It's one of the reasons Cynthia likes to keep writers and TV crews out of her lab. Which is a problem for me. Because if anyone can give me a clear steer on the future of intelligent machines, it's Cynthia Breazeal.

Before leaving London I'd spent several enjoyable hours sparring on the phone with the formidable but hugely likable Polly Guggenheim, Breazeal's PA and gatekeeper. Polly had informed me that the Personal Robots Group has a policy that Cynthia gives one ('and only one') face-to-face interview a year. 'We had the BBC here last year.' A pause. 'Never again. Now, who did I turn down last week? I've a list here somewhere. Would you like me to read it to you?' Polly speaks in a no-nonsense drawl built to come out of the mouth of a New York policewoman detective. Still, after eight conversations she hadn't said no, which gave me the hope I might be in with a chance. (Polly doesn't give the impression of being the sort of person to dally with niceties once she's made up her mind.) It was agreed that Cynthia would quickly interview *me* over the phone, to see if I was worth bothering with.

At that point I came straight out with it.

'Okay, seriously, what does it take to get this interview?'

'You really want to know?'

'Of course.'

'Dark chocolate. *Seriously good quality* dark chocolate.'

And so, after a brief chat with Cynthia, who sounded amiable and sparky, in which I promised to arrive at MIT bearing superior confectionery, I got my interview.

I find the Personal Robots Group on the fourth floor of the geek-famous MIT Media Lab, a boxy white structure on Ames Street that's beginning to look rather tired and shabby. Its residents will soon spread out into a shiny new ultramodern building next door, designed to encourage even more of the interdisciplinary collaboration that makes MIT such a hotbed of innovation. If there was a dial you could attach to a structure to measure the number of new ideas being generated inside, one attached to the Media Lab would probably blow a fuse before you could take a reading.

'One of the things that is so great about being at a place like MIT and the Media Lab,' Cynthia tells me later, 'is that we're not just a pure engineering group. We have skills and abilities that range from the aesthetic to the highly technical and scientific. That's unusual and it means we can answer questions that other labs find hard.' It's a fair point. Where else, I wonder, will you find research groups with titles like Affective Computing, Biomechatonics, eRationality, Information Ecology, Lifelong Kindergarten, Smart Cities, and, sharing the same lab space as Cynthia's Personal Robotics Group, Opera of the Future (this last looks at 'how musical composition, performance, and instrumentation can lead to innovative forms of expression, learning, and health'), all nestled up against one another? It's bonkers, in all the right ways. It's also an unholy mess.

Cynthia's lab reminds me of the scene in *Blade Runner* when genetic designer J. F. Sebastian returns to his apartment – a chaotic jumble of rooms, strewn with antiques and machine parts, and inhabited by comical pet creatures of his own design. As I enter, the plastic skeleton of a small robot sitting on a desk covered with computer parts raises its gaze from Ph.D. student Dan Stiehl to look at me. This is 'Huggable' (without his fur), a robot teddy made for small children and, Dan explains, designed to act either as a companion for when they are sick, or as an early education aid (where caregivers and teachers can control some of the robot's moves to encourage, for instance, the reading of a picture book). Even with his mechanical guts exposed, Huggable gives the definite impression of being alive. He tilts his head up slightly to gaze at me square on, waggles his ears and extends his robot paw. I laugh, a little nervously. It's spooky, but in a nice way. The cutting edge of robotics seems about as far away from *The Terminator* as you can get.

Dan shows me a 'Tofu' robot – an experiment to
toon animation techniques might be used as a way t
robots to express themselves. Tofu is 'a squash and stret
– and this one is a furry ball about the size of a grapefru
a small head dominated by wide eyes that dart comically
side to side and two feathery ears, along with a tiny orange be
that makes a cute chirping noise. Just like a cartoon character, Tofu can squish himself down and stretch himself out. He is laughably charming and if I were a child I would have immediately tried to take him home with me.

I'm eager to meet Leo, and Dan guides me behind a curtain. Suddenly I'm face to face with arguably the world's most lifelike machine. Except he's not very lifelike today. Leo is switched off. His head is slumped slightly forward and the box on which he sits has its front panel removed, revealing a myriad of strings and pulleys that control much of his movement. I also notice that Leo is wired up to sensors that hang from his curtained-off portion of the lab and, it turns out, no less than seventeen separate computers that process the information he takes in and generate his 'thoughts.' Leo is a prisoner of his limitations and, when you add in all those computers and sensors, actually much bigger than the cute pot-bellied furball I'd taken to be his entirety. It's disappointing, as Leo loses some of his personhood, and I'm a little crestfallen.

In person Cynthia Breazeal is like a shiny penny. She's forty-two, but looks at least ten years younger, a youthfulness no doubt related to her sporting prowess (an accomplished tennis player, she briefly considered taking up the game professionally). She manages to mix businesslike (a neat bob of a haircut) with childlike (her cheeks seem to permanently have a smile hanging off them, even when she's frowning). When she talks she uses her whole body. Her hands are constantly moving, as if literally grabbing concepts and shaping them in midair as she speaks. Her shoulders jog up and down, her head tilts with every new thought or idea, and she seems eager to laugh. People might argue about whether her machines are alive or not, but there's no debate about Cynthia. She's about the most alive person I think I've ever met.

'These are good,' she says tucking into the chocolates with a broad smile. 'You should have one!'

I decline, wanting all that chocolatey goodwill going into our interview. Plus I know how much they cost, and I'm not sure my tastes are expensive enough to do them justice.

Polly Guggenheim interjects, 'It'll do your endorphins good.'

'Antioxidants! Antioxidants!' exclaims Cynthia.

'Exactly! It's like red wine!' says Polly as she departs.

With the benefits of chocolate and red wine firmly established I tell Cynthia how affected I was watching videos of Leo, especially in comparison to dozens of other robots I'd viewed for my research. Here was something of the future of robots long imagined by science-fiction writers but that I'd never seen for real. Leo *grabbed* me (and indeed all the friends I showed him to, some of whom refused to believe he wasn't a puppet). I believed he was thinking. The tilt of the head, the blinking, the shrugging of the shoulders are all social cues that I naturally understood and responded to, making me attribute emotional states and thought processes to him. I gave him a personality. In fact, I couldn't help myself, because this was a robot acting in a, well, recognisably human way (and therefore the exact opposite of Keanu Reeves). I'd had a similar reaction to Huggable only minutes earlier. What, I wonder, are Cynthia and her team doing right?

'I think it's because we put *people* at the core of what we're trying to do,' explains Cynthia. 'A lot of work in robotics is still very focused on technology, but in our lab we put these robots in front of real people so we can understand their impact. My group takes the relationship between social robots and people *seriously* and we're trying to design for *both sides*. It's not just about having the robot understand people, we're trying to make people *understand the robot* so you're naturally able to use your own way of thinking about the world to understand what must be going on in the ...' She pauses. 'Er, mind, so to speak, of the robot ...'

'So to speak?'

'Well, it's a loaded term, isn't it?'

Two minutes in and we've already hit a key problem that's dogged Artificial Intelligence since day one, namely the intellectual quicksand around definitions of 'mind' and 'intelligence.'

'Do you think Leonardo thinks?' I ask.

Cynthia does a neat sidestep. 'I can ask a range of philosophers that question and they'll all come up with a different answer.' But I push her. If it's robot-minds rather than robot-bodies that have been holding our machine brethren back, and if she believes Leo *is* genuinely thinking, that's profound.

'You're kind of avoiding the question,' I say. She laughs and considers for a moment.

'What Leonardo is able to do is *aspects* of what I describe as thinking.' And then she's off, a rapid-fire walkthrough of Leo's major subsystems, about how he takes in 'perceptual information,' how he has 'internal states,' and how his 'cognitive-affective circuits' are 'influencing internal actions,' 'propagating that activation' which 'ultimately boils down to physical action in the world.' She takes a breath. 'I mean it's *all there* in some sense, but it's at a much simpler and rudimentary level, I think, than what's happening in animals and people.' So, *is* Leo thinking? Well, sort of. A bit.

Part of the problem with this sort of question is that it's human beings that ask it, and we have a natural prejudice. We model the concept 'thinking' on how *we* perceive ourselves thinking. Breazeal's quite clear that robot thinking *won't* be the same as human thinking, just as robot emotions won't be the same as human emotions.

'When we talk about emotions, it's not about whether a robot *can* have emotions but what are robot emotions? So dogs have dog emotions, they don't have human emotions. But owners have beliefs about what their dog must understand about them, but they really don't *know*, right? They project their own emotions onto the dog!' She laughs. 'But they see the dog's emotions as being genuine for what a dog is, and they see the relationship as being genuine, and that's the point. I want people to accept robots as genuine personalities, so they can have an authentic relationship with them.' So, *does* Leo have emotions? Well, sort of. A bit.

'There's a lot of confusion in the media who often seem to think "you're going to replace people, you're trying to make *people-equivalent* machines." I'm like, "No! These are *robots*." But they misreport this stuff. They don't get it!'

The question of comparing human and artificial intelligences was addressed in a 1984 lecture by computer science genius Edsger Dijkstra, who said that the question 'is about as relevant as the question of whether submarines can swim' – a warning against trying to come up with a human-centric definition of 'thinking.'

Cynthia makes the point that different intelligences are a function of the bodies they're born into. 'So the question is, what kinds of bodies are critical for developing what kinds of minds? Dolphin intelligence isn't human intelligence for a reason. Cat intelligence isn't dog intelligence for a reason. I would never say that what we're doing with these robots is somehow explaining what humans are really doing.'

In 1997, IBM's computer Deep Blue beat chess grandmaster Garry Kasparov. This is seen, rightly, as a landmark moment in AI research, but it's part of an old AI tradition of trying to get machines to replicate highly intellectual 'symbolic' reasoning (the sort you need to master algebra, for instance, and the kind of tasks revered, says Rodney Brooks, by 'highly educated male scientists'). It's no coincidence that the frontiers of AI research seem to have been populated with computers that get ever better at playing chess. But the big elephant in the room for this kind of AI research is that Elephants Don't Play Chess (this being the title of a paper Brooks wrote in 1990), but they do do any number of 'common-sense' things that machines can't – as can dogs, cats and preschool children. Sure Deep Blue can occasionally beat grandmasters (and no doubt elephants) at chess, but it'll never be able to find shade or express its political views on Vladimir Putin.

So, does Deep Blue think? Well, sort of. A bit. But we still haven't got anything like *Star Wars*' C-3PO, the multipurpose smart machine that we consider a real personality. Or as Steven Pinker, Johnstone Professor of Psychology at Harvard University, wrote:

> The main lesson of thirty-five years of AI research is that the hard problems are easy and the easy problems are hard. The mental abilities of a four-year-old that we take for granted – recognising a face, lifting a pencil, walking across a room, answering a question – in fact solve some of the hardest engineering problems ever conceived.

Brooks argues that 'simple things to do with perception and mobility [are] a *necessary basis* for higher-level intellect.' In other words, if you want to build something truly smart it needs to be in the world, and not chained to a chessboard. (Garry Kasparov's ability to play chess, after all, isn't isolated from the rest of Garry Kasparov.) In the parlance of Brooks and co., it needs to be 'embodied' and 'situated,' meaning it needs a physical body via which it senses the world and it is this, more than abstract concepts, that should drive its actions.

'Sense and action. This is all I would build, and completely leave out what traditionally was thought of as the intelligence of an artificial intelligence,' wrote Brooks in *Robot*. Using this approach he soon began to achieve impressive, lifelike results from his machines, much to the chagrin of the AI old guard. At the Second International Symposium of Robotics, held just north of Paris in 1985, Brooks demonstrated Allen, a robot that could autonomously navigate itself down corridors, avoid pedestrians and choose a route based on the obstacles before it – all using a 'sense and react' approach.

This was a marked contrast to previous attempts at robot locomotion that relied on robots trying to build internal 'mental maps' of their surroundings and referring to these internal representations, rather than directly to the world. Time and again these 'symbolic' models couldn't keep up with an ever-changing world. Brooks recalls how strong shadows had foxed

Hans Moravec's attempts to achieve autonomous motion in an MIT robot called the Cart. 'The Cart got very confused about its model of the world, and Hans had to run out and fix the world occasionally so that the internal model was a little more accurate – it was easier to change the real world than to change the Cart's internal model of that world.' Allen by comparison was, as Brooks puts it, 'glorious.'

By 'eliminating the process of reasoning' and making a machine 'completely based on unthinking activity' that made a 'direct connection of perception to action,' he had built something that seemed to act much *more* intelligently. In animals, 'complex capabilities are built on top of simpler capabilities' he argued – and started building his robots accordingly. It was a view that seemed way too, well, *simple* for those 'highly educated male scientists,' and Brooks was soon working in his favourite place – on the fringes of respectability.

Step forward two decades and Brooks has largely won the debate. You can find footage of the aforementioned 'BigDog' on YouTube traversing inclines covered in soft soil or woodland debris, snow-covered slopes, even piles of rubble. There are two particularly startling moments in these videos. In one, BigDog begins to slip on some ice but is able to recover and remain upright. In another, it's given an almighty kick from one side. BigDog staggers but quickly regains its footing. It's not chess, but suddenly this machine looks *conscious*. It's Brooks's 'sense and action' argument taken to a whole new level.

Cynthia's work carries on in the Brooks tradition. Step one: build a body. Step two: make it sense and react to the world around it. But it's the next step where Cynthia comes into her own. *Make it social*. Or to put it another way, make some of its reactions 'social' reactions.

So, for example, if someone gets too close to the robot, not only does it back away, it might express its annoyance at having its personal space invaded with a grimace or sound of disapproval. In these examples the robot responds directly to something you *do*. But Cynthia is striving for a robotic social intelligence that takes things into a whole new realm, responding to something you *think*, where the robot perceives something of your mental state from the way you are acting, perhaps understanding if you are sad, happy or confused – or working out that you believe something different to it. She admits we're a long way off achieving anything that looks like this in a robot although in certain highly controlled conditions Leo has been able to work out that a researcher believes an object is in one box, even though Leo knows it's in another, a version of something psychologists call the 'false belief test'.

Having the robot understand you, your actions and your mental states is only one half of the interaction. As Cynthia said, it's not just about having the robot understand people, it's helping people understand the robot. 'It's about getting robots to understand, connect and relate their mental state and yours – so you can work together in a normal, social way,' says Cynthia, 'You "get" each other.'

This is because the Personal Robotics Group is trying to understand not only how to design a robot that can learn, but how you design a robot that's teachable by a person who knows nothing about robotics, AI or machine learning. 'The challenge we're addressing is, how do you allow Grandmother to teach the robot? The world is full of people, so social forms of learning – observation, imitation, tutelage – all these things are really important. We're one of the few groups in the world that has robots learning from people who know nothing about robotics.' She laughs. 'And I've gotta tell you, it's not easy to design a robot that can learn from people, because people do all kinds of stuff!'

Of course, as humans we learn to deal with all the 'stuff' that other people do as we are growing up, and are hopefully nurtured by our parents and caregivers. Children generally develop in a safe and constrained environment where they can make mistakes and their brains can mature. This is all part of becoming intelligent in a way that means you can read a book, do your job, realise if someone is upset, drive a car and maybe (just maybe) understand cricket. After all, Garry Kasparov didn't pop out of the womb with a detailed knowledge of the Petrov Defence. First he had to learn how to walk, talk, eat and 'generalise task goals to new configurations' just like the rest of us.

'Look at how often children fall when they're learning to walk,' says Breazeal. 'I mean, it's like *huge*. It's not until you start really digging into all this that you start to realise how deep and profound and *difficult* this kind of stuff is to do. Then you start to appreciate how amazing it is that *people* can actually do it!' She's laughing a lot now. 'You just take it for granted that a smile can be interpreted as bemusement, or joy, or any number of things. It's obvious. You just get it.' She raises her eyes and holds open her arms in an expression that (rather demonstrating her point) I can instantly interpret. *Holy Moly!*

'Now!' exclaims Cynthia, 'you try to build a robot that can do that. *Oh my god!*'

Take computed attempts to capture the nuance of 'natural language' (the ability to use words in a conversational style). One popular example is Eliza, developed by MIT's Joe Weizenbaum in 1965. (Eliza was named after Eliza Doolittle from the George Bernard Shaw play *Pygmalion*, in which professor of phonetics Henry Higgins teaches the East End flower girl to imitate the speech patterns of high society.) Weizenbaum tried to get Eliza to mimic a technique used by many psychotherapists – reflection. Reflection is essentially a 'mirroring' or restating of what has just been said and lets clients know they are heard, hopefully encouraging them to open up more. So, if you say, 'I want to kill my husband!' they might respond by saying something like, 'Your husband is annoying you?'

You can find various copies of Eliza on the Web. It's surprisingly good. In fact, it compares favourably to other natural language programs developed much later (highlighting again how hard it can be to improve the performance of AI in areas that humans take for granted). As part of my research I decide to get some online therapy.

I start off by typing, 'I'm writing a book.'

'How long have you been writing a book?' inquires Eliza.

'A few months,' I respond.

'Tell me more,' she counters, as if genuinely interested.

'I'm trying to work out what the future might be like.'

Then she gets heavy.

'Do you believe it is normal to be trying to work out what the future might be like?'

I have to say this makes me think. Eliza really is turning out to be like a therapist. But then it all starts to go a bit weird. After due consideration I conclude that yes, actually it is perfectly normal to try and work out what the future might be like and reply simply: 'Yes.'

Eliza's response: 'I see.'

Even though Eliza's a piece of code running in my browser, this still feels like an awkward silence. I'm uncomfortable. 'You don't do well at parties, do you?' I type.

'Oh . . . don't do well at parties do me?' says Eliza and I'm not entirely sure what she means.

'Are you asking me to do you?' I reply. 'Maybe you do better at parties than I thought!'

'Why you interested in whether or not I am asking you to do me?' she says. 'Maybe me do better at parties than you thought?'

I can't work out if Eliza thinks I'm a pervert or is a machine impersonation of a nymphomaniac foreigner. Poor Eliza, she failed to get the irony in 'You don't do well at parties, do you?' which led us somehow into a dodgy 1970s comedy script.

It's results like this which lead AI and robotics researchers to suggest that the best approach to building intelligence is not to try and make an 'off the shelf' mind, but to grow one in the robot, to bring it up. Children slowly learn not to make mistakes like Eliza's over years and years of interacting with those around them. 'Natural language' skill takes *thousands of hours* to master (and we can all still get it wrong, as any 'misunderestimated' president will tell you).

Perhaps if AI research had taken note of one of its founding fathers, it may have saved some time. In a landmark and it turns out startlingly prescient paper, written in 1950 (in which the first line was aptly, 'I propose to consider the question, "Can machines think?"'), the great, tragic, war-shortening Alan Turing wrote, 'Instead of trying to produce a program to simulate the adult mind, why not rather try to produce one which simulates the child's? If this were then subjected to an appropriate course of education one would obtain the adult brain.'

This proposed necessity of having to raise robots might lead you to the conclusion that truly intelligent robots will be few and far between. But the thing about robots is you can replicate them. Once we've got one intelligent robot brain, we can copy it to another machine, and another, and another. The robots have finally arrived, bringing an explosion of 'strong AI'. Of course, it may not just be us (the humans) doing the copying, it might be the robots themselves.

And because technology improves at a startling rate (way faster than biological evolution), one has to consider the possibility that things won't stop there. Once we achieve a robot with human-level (if not human-like) intelligence, it won't be very long until robot cognition outstrips the human mind – marrying the human-like intelligence with instant recall, flawless memory and the number-crunching ability of Deep Blue.

Cynthia may want to build a human-robot team, but what happens when one half of that team (the human) is starting to look like the village idiot? This is what has been called 'The Singularity' – the moment where the generation of new knowledge (through a merger of human-like talents of imagination, curiosity and creativity with machine-like computational muscle) starts to look like a rocket leaving a launch pad. According to The Singularity's prophets (notably Ray Kurzweil), when this happens there is only one strategy open to us: we must merge with our machines.

If you can't beat 'em, join 'em.

I think back to something George Church said: that one way of looking at the human being (and therefore the human brain) is 'simply' as a collection of unthinking tiny bio-machines computing away – reading genetic code and spewing out 'computed' proteins and the rest. We're machines too, just 'wet' biological ones. Brooks also makes this argument:

> The body, this mass of biomolecules, is a machine that acts according to a set of specified rules ... Needless to say, many people bristle at the use of the word 'machine.' They will accept some description of themselves as collections of components that are governed by rules of interaction, and with no component beyond what can be understood with mathematics, physics and chemistry. But that to me is the essence of what a machine is, and I have chosen to use that word to perhaps brutalise the reader a little.

Intelligence and consciousness are computable, because you and I are computing it right now. I compute, therefore I am.

George Church, though, put across a less brutal take on the human machine. 'I think of us more and more as mechanisms,' he told me. 'We're starting to see more and more of the mechanism exposed and it just makes it more impressive to me, not less. If someone showed me a really intricate clock or computer that had emotions and self-awareness and spirituality and so forth I'd be very, very impressed, and I think that's where we are heading, where we can be impressed by the mechanism.'

One thing is certain. If a conscious human-like intelligence *is* 'computable' by a machine, the processing power to compute it will be within reach of the desktop very soon. Hans Moravec wondered 'what processing rate would be necessary to yield performance on par with the human brain?' and came up with the gargantuan figure of one hundred trillion instructions per second.* To put this number in context, as I was ushered into the world in the early seventies, IBM introduced a computer that could perform one million instructions per second. This

*A hundred trillion looks like this: 100,000,000,000,000. Moravec reckons our brains do this number of discrete calculations *a second*. As Joey said in *Friends*, pointing to his head, 'It's not just a hat rack.'

is *one millionth* of Moravec's figure. As I write, Intel has released its 'Core i7 Extreme' chip, which is over *one hundred and forty thousand times* faster – or about one seventh of Moravec's figure. At this rate, your new laptop will achieve the same computational speed as the human brain before the decade is out. Soon after that, if the exponential trend continues, your laptop (or whatever replaces it) will have more hard-processing muscle than all human brains put together. This will happen sometime around the middle of the century, according to Ray Kurzweil.

Supercomputers have passed Moravec's milestone and it's therefore no surprise to find various projects using them to try to simulate parts of animal and human brains, merging neuroscience and computer science in an attempt to get to the bottom of what's really going on in that skull of yours. Henry Markram, leader of the Blue Brain project (which works by simulating individual brain cells on different processors and then linking them together) believes 'It is not impossible to build a brain, and we can do it in ten years.' He's even joked (or not, depending on how seriously you take the claim) he'll bring the result to talk at conferences. At the 2009 *Science Beyond Fiction* event in Prague, Markram told his audience that he believes 'the ethereal "emergent properties" that characterise human thought will, step by step, make themselves apparent.' In other words, by simulating the brain at finer and finer levels of granularity, his project will finally allow us to perceive what makes consciousness work – it will reveal the laws of intelligence and cognition and we will be able to use them as readily as the laws of motion or computation today. It's no surprise to find that lots of people disagree with him.

But something's not sitting right with me, and it's not that I don't like being called a 'machine'. In fact, the machine metaphor kind of makes sense given what I found out at Harvard. I can't quite place my finger on it so I ask Cynthia her view on throwing processing power at a simulation of the human brain as a way to create an artificial intelligence.

She does a tiny little snort.

'The bottom line is there's still a long way to go before we can have that simulation actually do anything. I mean they can run the simulation but what is it doing that can be seen as being intelligent? How does that grind out into real behaviour, where you show it something and have it respond to it? There's a lot of understanding that needs to be done. We're making fantastic strides but I think,' she drops to a conspiratorial whisper, smiling, 'there's a lot we still don't know!'

Cynthia has nailed the root of my discomfort. Someone can give you the best calculator in the shop, but if you've never learned any maths, it's largely useless to you. If the brain is computable, it's not that we won't have the processing power to recreate its mechanisms, but that we're still a long way off working out how to drive that simulation. If you'd never learned to read, your eyes could take in the shape of every letter on this page, but it would mean nothing to you, and photocopying it a hundred times (or even inventing the photocopier in order to do so) wouldn't help you either. Just as you had to learn to read, AI and neuroscience research collectively have to tease out not only what it is they're looking at, but what it means.

Sure, there's exponential growth in processing power, but the jury is out as to whether there is an equivalent growth in understanding how to use that power more 'intelligently,' to create (to paraphrase one of Henry Markram's analogies) a concerto of the mind by playing the grand piano of the brain. If there had been this growth, maybe your new laptop would be one-seventh as smart as you are. But it isn't. This is where the strength of projects like the Blue Brain (and Cynthia's work) really lies – as tools to slowly help us pose the right questions that will lead to a better understanding of intelligence, emotion and consciousness.

'The thing that's useful about trying to build sociable robots is it pushes us to ask those questions,' says Cynthia. 'What is life? What is mind? Robots push us to be more precise. They push us to really think hard about what is it that we really mean by those terms.' The answers, she believes, are 'decades off.' In fact, she's very careful to dampen expectations about her own research. I ask her, 'What's the longest interaction you've had with a robot?'

'Where it was actually meaningful?'

She pauses for a long time. 'It's in the order of like. . . .' She hesitates.' . . . tens of minutes, you know? It's short.'

Leo's never getting off that box.

It seems to me that the powerhouse that drives and defines real intelligence is curiosity. It's why Cynthia's at MIT, why I'm in her office, why you're reading this book, and why a man called Hod Lipson at Cornell University may have just done something profound. Because he's built a machine that seems to exhibit genuine curiosity – a machine that is already working with its creators to generate new knowledge. Accordingly I make Hod my next port of call.

Before I get to Cornell, though, I have two tasks. One is a short gig at the friendly Mottley's Comedy Club in Chatham Street, and the other is to come up with a new word for 'robot.' I'd asked Cynthia if she thought the word was restrictive, given all its associations in popular culture. Did it maybe fail to get across her idea of a wide range of sociable machines with their own take on emotions? 'Yes, I'd like to come up with something different,' she'd said, and then looked me straight in the eye. 'How about you come up with a better term? And I'll start using it!'

The word I come up with, on my journey upstate, is 'AInimal' (pronounced 'ay-nimal' or 'ay-eye-nimal' depending on your preference). A creature with an artificial mind, but a creature nonetheless.

CHAPTER 5

Like explaining Shakespeare to a dog

I have no special talents. I am only passionately curious. ALBERT EINSTEIN

How do you find a truth? Here's one: some students at Cornell University smoke weed. How do I know this? Because shortly after arriving in the small town of Ithaca I find myself standing outside *Insomnia Cookies*. 'Warm Cookies Delivered Late Night' says the purple neon sign in the window. I don't smoke but I've been around enough fans of the herb to know that they tend to get a craving for baked snacks ('the munchies'). This makes putting a late-night warm cookie store (that will deliver!) on a university campus nothing short of *retail genius*. I subsequently discover that *Insomnia Cookies* has outlets on eighteen campuses throughout the States.

It is this kind of creative thinking that AInimals find hard. It's also hard for them to infer a wider truth from limited data. There isn't an AI in the world that has spent one night too many listening to stoned, cookie-eating students spouting cod-psychology to a soundtrack of space rock – and so there's not one that could, given sight of a single cookie store, make the deduction I've been able to: late-night campus cookie store = 'the munchies' = stoned engineering students.

That said, sitting just a few blocks from where I'm standing is a computer brain that *can* find truth from looking at slivers of data, using previous experience to make its deductions. It's a brain born of the body now doing some serious intellectual work – a 'sense and action' robot that has graduated.

I find one of the machine's creators, Professor Hod Lipson, in his lab on the second floor of Upson Hall, a nondescript university building just off the 'Engineering Quad' of Cornell's verdant campus. He has a round, solid frame, thick curly black hair and a big smile. If Tom Hanks had become a powerlifter instead of an actor, he'd probably look a bit like Lipson. While most of the scientists I've met are driven by an almost insatiable curiosity, Lipson takes curiosity to a new level, literally. He's curious about curiosity.

'Artificial intelligence is a moving target,' Hod asserts, taking a seat opposite a table piled high with AI textbooks and professorial paperwork. 'I want to create something nobody can argue *isn't* intelligent. I was thinking, what's an unequivocal hallmark of intelligence? I think it's creativity, and particularly *curiosity*.'

This is why I've come to see Hod. If our future is to include truly intelligent machines then they'll need to be curious and creative ones. If we can produce such AInimals they could radically reshape how we go about finding and applying new knowledge, heralding a revolution in thinking similar to the one commercial robots have already achieved in manufacture. I get straight to the point. 'Does a curious and creative machine mean a sentient machine?'

'Well, what does that mean?' asks Hod. 'I have to push you on what you mean by "sentient."'

Bollocks. Less than a minute in and I'm getting strong *déjà vu*. Cynthia tried to sidestep the question of what constitutes machine consciousness. Hod wants *me* to answer it, which seems a bit unfair given he's the professor. I do the only thing I can. 'Well, let me ask you,' I say. 'What do *you* mean by it?'

Hod pauses. I'm not sure he was expecting a return serve, especially one that in any decent rulebook would be considered cheating.

'I interpret it as deliberate versus reactive. Er ... human-like . . .' He pauses. 'I don't know.'

I claim a draw.

Lipson's lab became famous (in robotic circles) for its work building AInimals that are arguably *self-aware*. The lab's Starfish robot built by Hod, Josh Bongard and Victor Zykov is notable for learning to walk from first principles. (Just like Leo, the Starfish is actually sitting inert, almost forgotten on a shelf in the lab, when I visit). It wasn't programmed to walk; instead, Lipson and his team programmed it to learn about itself. It then used *this* knowledge to work out how to move.

Just like a baby, the Starfish developed a self-image through physical experimentation. Hod explains, 'It moved its motors, it sensed its motion and then it created various models of what it might look like – "maybe I'm a snake? maybe I'm a spider?" We told it to create multiple explanations of what it knows so far.' The robot then tested each model by putting them in competition with each other.

'We tell it to create a new experiment that creates disagreement between the predictions,' says Lipson.

One of the key strengths of human intelligence is its ability to contradict itself. Our capacity to hold conflicting ideas allows us to argue things out in our heads and collectively as a society. What Hod and his team did was build this productive contradiction into the Starfish. 'We set off two lines of inquiry. So one of them is the thing that creates models and the other is the thing that asks questions. They have a predator-stalks-prey relationship. The questions basically try to break the models.'

Over time the predictions of the competing models get closer to reality until one model is deemed sufficient for the robot to say, 'This is what I look like.'

It has to be said that the Starfish isn't the most graceful creature in the world. It doesn't so much walk as stagger and flop forward. Less Ginger Rogers, more gin and tonic. Still, the achievement is not to be sniffed at. This was an AInimal actively learning to do something no one had taught it to. But what happened next is why I've come to Cornell. The team's AInimal brain, born of physicality, started to do some extraordinary things in the symbolic arena (you know, all that 'educated male scientists' stuff). It started to churn out scientific truths.

One could argue the search for truth is history's guiding force. There have been libraries of books written (and lost) on the subject. But if someone said to me, 'Go on then, history of truth in five minutes,' I'd probably reach for two key figures – Socrates (born Greece, 469 BCE) and Francis Bacon (born England, 1561), not least because they both died in interesting ways.

Socrates was put to death by the state of Athens for 'refusing to recognise the gods recognised by the state' and 'corrupting the youth'. Despite the opportunity to escape his fate, the philosopher placidly took a draft of poison hemlock prepared by the authorities. Francis Bacon, many believe, died as a result of trying to freeze a chicken. It might seem odd, therefore, that both are seen as key figures in the history of *reason*.

But both Socrates and Bacon were very good at asking useful questions. In fact, Socrates is largely credited with coming up with a *way* of asking questions, 'the Socratic method,' which itself is at the core of the 'scientific method,' popularised by Bacon during the Enlightenment – a period of European history when 'evidence' and 'faith' had an almighty bun fight and the balance of power between church, state and citizen was questioned as philosophers and scientists challenged the prevailing orthodoxy of religious authority.

The Socratic method disproves arguments by finding exceptions to them, and can therefore lead your opponent to a point where they admit something that contradicts their original position. It's powerful because it can get people to admit to themselves that they're wrong. It's pretty good at exposing your own (as well as others') prejudices and gaps in reasoning. Lawyers use it a lot (though don't let this colour your opinion of it).

By the time Francis Bacon went to university, the teachings of one of Socrates' students, Aristotle, had become the accepted way to conduct 'scientific inquiry.' Aristotle had pioneered 'deductive reason,' the practice of deriving new knowledge from foundational truths, or 'axioms.' It was generally believed that if you got enough boffins together to have a solid debate, scientific truth would be teased out over time.

This have-a-long-and-involved-chat method worked well for mathematics and still does, because the axioms (the basic mathematical operations – plus, minus, divide, multiply) are long established. But it wasn't as useful for finding out new knowledge relating to the physical world. Much to Bacon's dismay it seemed that 'science' involved sitting around and talking. Nobody was getting off their arse and observing anything new or doing any experiments.

In common with Socrates, Bacon stressed it was as important to disprove a theory as it was to prove one – and real-world observation and experimentation were key to achieving both aims. Bacon also saw science as a collaborative affair, with scientists working together, challenging each other. All these remain hallmarks of good scientific practice today – observe, theorise, experiment ... and then try to prove yourself wrong – all in collaboration with peers who can (and should) give you a hard time. Bacon himself wasn't a distinguished scientist; his main contribution was to describe and champion an evidence-based scientific method. That said, he did do the odd experiment, including the one that killed him.

While travelling from London to Highgate with the king's personal physician, Bacon wondered whether ice might be used to preserve meat. The two men got off their coach, purchased a chicken and stuffed it with snow to test the theory. In his last letter, Bacon is said to have written, 'As for the experiment itself, it succeeded excellently well'; but the act of stuffing the chicken led to him contracting fatal pneumonia, killing him a few days later. Possibly the only instance of Bacon being killed by eggs.

Back in modern-day Ithaca, the spirit of Socrates and Bacon were coming out of a robot.

Hod (along with colleague Michael Schmidt) wondered if the 'sense and action' brain from the Starfish could go beyond working out what its own body looked like and begin to reach useful conclusions about the wider world. To put it another way, having gone through the process of building an accurate mental model of *itself*, could it make accurate mental models of other things?

Their first idea was to give the robot brain the ability to set up the starting position of a 'double pendulum' before letting it fall. It was then fed the results of each experiment – which had been recorded using motion-capture technology – allowing it to accurately assess the pendulum's motion.

A double pendulum is basically one pendulum with another pendulum stuck to the bottom of it – two sticks joined end to end by a free-moving hinge. While the top half swings from left to right, the bottom half soon starts to go a little crazy. Because the bottom section isn't attached to a stationary point (like the top pendulum) but to something that's already in motion (the lower end of that swinging top pendulum), it will swing left, swing right, spin round clockwise, or counterclockwise, seemingly at random. Lipson and Schmidt chose the double pendulum because it's a good example of a system that is simple to set up but quickly exhibits chaotic behaviour – and would therefore be a good test of the robot brain's skills.

The results were startling. In fact, the Starfish brain went a long way to deriving the laws of motion, previously sweated over for decades by the likes of Isaac Newton. And it did it in three hours. It followed the same process as when it sat in the Starfish, guessing at models (mathematical equations) that might explain what it had seen so far, then creating new experiments (new starting positions for the pendulum) that targeted the areas of most disagreement between equations it had guessed at so far. 'With the double pendulum it very quickly puts it up exactly upright, because some models say it's going to fall left and some models say it's going to fall right. There's disagreement. It's not a passive algorithm that sits back, watching,' says Hod, smiling. 'It *asks questions*. That's curiosity.'

One of the key things this AInimal was curious about was 'invariants'. Or, as Hod explains, 'Let's say you have two sets of measurements coming in from some experiment and they wiggle up and down in some complicated way. Our machine tries to find out what's *constant* about them. Maybe if you add the two figures together, you always get the same number? Maybe it's more complex – perhaps the sum of their square roots is the thing that stays the same.'

These hidden 'invariants' often provide crucial insights into what's guiding a system's behaviour. 'Asking what stays constant about the pendulum can help you discover that the force you apply to something is always equal to its mass multiplied by its acceleration; Newton's Second Law of Motion.'

Speaking with Hod, I see a hemlock-taking philosopher and a chicken-freezing experimentalist partially reincarnated in machine form. The robot's programming consigns inaccurate models to the dustbin by getting it to admit that others offer a better explanation of the real world (hello, Socrates) with evidence won via experimentation (hello, Bacon), sniffing out fundamental, unshifting constants hidden in the data. Lipson has created a computer-based way of asking good questions – one of the hallmarks of true intelligence.

However, we're still a long way off what Hod (or anyone else) would call a truly intelligent machine. The machine didn't know it had found laws of motion; it took Hod and his colleagues to recognise the equations. 'A human still needs to give words and interpretation to laws found by the computer,' says Schmidt.

Without meaning the results are largely worthless (which is not to say the time saved is not incredibly useful). I'm reminded of the classic scene in *The Hitchhiker's Guide to the Galaxy* in which a hypercomputer called Deep Thought is built by a race of super-smart humanoids to compute the answer to the ultimate question of 'life, the universe and everything', which after millions of years of deliberation it announces to be '42', before suggesting that the humanoids don't know what 'The Question' is.

No one understands the irony in this story more than Hod Lipson. He set his machine looking at a process within soil bacterium. True to form, the program generated an equation in double-quick time. But what did it mean?

'We're still looking at it,' says Hod with a smile. 'We're staring at it very intently. But we still don't have an explanation. And we can't publish until we know what it is.'

'You don't understand what it's saying?'

'No,' says Hod.

'It's your 42 moment?'

'Yeah,' says Professor Lipson, smiling ruefully.

Several months after my visit I email Hod to see if he's got anywhere with the mystery equation. 'We're struggling,' he replies. 'Maybe it's hopeless. Like explaining Shakespeare to a dog.'

Even so, the machine's achievement is important. 'We can go from data straight to laws, whereas previously people could only go from data to predictions. So now a scientist can throw it some data, go and have a cup of coffee, come back and see fifteen different models that might explain what is going on. That saves a lot of time. Previously coming up with a predictive model could take a whole career. Now at least you can automate that so you can focus on meaning.' That's a powerful enabling technology, because it leaves more time to think. More time to do that thing that humans are good at and computers aren't – attributing meaning to things. Hod is doing for thinking what dishwashers have done for after-dinner conversation.

I ask him if his brain could come up with a model of how to learn. After all, if it can come up with a model of itself, or a model of energy conservation derived from looking at a pendulum, why not get it to observe something learning and see if there are insights that could be applied to machine learning and the further development of AI? Hod laughs. 'That's what we're working on now. We're working on what we call 'self-reflective systems' where they are thinking about thinking.'

Hod has clear reasons for doing this. 'If you want to get to human-like intelligence, you need a brain that can think about thinking. That's self-reflection which is important in life. You can learn things the hard way, or you can think about how you've been thinking.'

Curiosity? Self-reflection? Social interaction? Thinking about thinking? While AI still has a long way to go, the questions researchers like Hod and Cynthia are raising – and the results they are getting – are significant. It's clear that machine intelligence (both 'social' and 'symbolic') is advancing, if in very small increments. The question is, how might this impact you and me? How might it all play out?

One option is that AI and robotics continue to make advances but are unable to achieve a human-like or human-level intelligence: that consciousness is 'non-computable,' or even if it is computable the human brain may, ironically, not be smart enough to work it out. In this scenario, machines like Hod's will keep churning out models and equations, but it'll still be humans making (or failing to make) sense of them. The rate at which knowledge advances will go up a good few notches, aided by ever more powerful technology, but the ultimate speed limit will be how quickly we can take meaning from what our machines are helping us discover. (Shortly after our meeting, Hod put the code for his brain – named 'Eureqa' – online, for anyone to feed data into.)

Cynthia's robots, meantime, will get ever more social, and her research will help us interface with our technology in ways more natural to us. But this sociability will be a result of some impressive computer processing that falls short of machine consciousness.

As legendary AI researcher Marvin Minsky remarked upon seeing Leo, 'My objection to Leonardo is, it's just a trick. It doesn't really have emotions. It just knows how to fool you into thinking it does.' But even if Minsky is right, it still doesn't take away from Cynthia's key achievement to date: that we can sometimes *accept* her robots as emotional and sentient, even if they're not.*

Option two is where machines *do* achieve a level of intelligence we accept as sentient, something that goes beyond trickery. Somewhere on this road, we'll have to deal with the sticky issue of whether robots have, er, human rights. As Nick Bostrom, who I met in Oxford points out, 'Whether somebody is implemented on silicon or biological tissue – if it does not affect functionality or consciousness – is of no moral significance. Carbon-chauvinism is objectionable on the same grounds as racism.' People complain about robots taking over their jobs now. Imagine the reaction when a robot can sue for unfair dismissal on the basis of 'machinism.'

If the AInimals do get truly smart, we might enjoy a confederacy of man and machine. We remain separate but work together. This is a world where sometimes our machines 'explain things to us,' says Hod. Some machines are good (C-3PO) and some machines are questionable (HAL 9000). Machines become part of humanity. They're 'us' as much as the French and English are 'us' to each other – foreign, sometimes hard to get along with, at other times the people we'd prefer to spend our holidays with but in the final analysis a part of our global society.

But there are, of course, more apocalyptic visions of this future. After the machines get smart, they engineer themselves

*Cynthia says Minsky is 'missing the point' by 'only focusing on the internal aspects of emotion and not the interpersonal and using human emotion as the gold standard. Dog owners don't say their dog's emotions are a "trick" or "fake" even though they are not human emotions.' On a return trip to Boston to see Ray Kurzweil, I pop back into the Personal Robots Group lab and see its latest sociable robot, Nexi, in action. In tune with the lab's focus on human–robot relationships, Nexi is interacting with a boy no older than eight who is enthralled by her human-like tracking of his movements as he dances in front of her. Despite being made of moulded white plastic Nexi's face expresses a whole gamut of emotions – her big eyes blinking, her white plastic 'eyebrows' moving, her mouth ranging from slack-jawed boredom (when not much is happening) to tight-lipped interest (if the boy is active) or annoyance (if the boy gets too close). It's startling to see how quickly both the boy and those of us watching accept Nexi as somehow sentient and alive.

to get smarter still. After all, we'll have worked out how to 'compute' intelligence and the AIminals, benefiting from the continuing spiral in processing power, will compute that intelligence faster than we can. Within a few short technology generations, they could leave us behind in a cognitive backwater called 'humanity.' We become to our AInimal progeny what primates are to us. At this point we hope they feel some responsibility for their ancestors and don't break us down for raw materials, try to hunt us to extinction (*Terminator*) or (yes, *Matrix* fans) use us for batteries. They rule us, and our fate will be in their hands.

Option three is that man and machine merge. As I found out during my chat with Nick Bostrom, this is already happening. In the opening pages of *Robot*, Rodney Brooks writes, 'Recently, I was confronted with a researcher in our lab, a double-leg amputee, stepping off the elevator that I was waiting for. From the knees up he was all human, from the knees down he was robot.' In late 2009, Robin af Ekenstam became the first recipient of 'SmartHand' – a prosthetic mechanical hand that not only replicates the movement of its human equivalent, but also provides Robin with a sense of touch (the replacement hand is wired to existing nerve endings in the stump left after he lost his original hand to cancer).

It's not a huge leap of the imagination from mechanical body parts to a world in which we also merge with mechanical brains. This future comes in two flavours. The 'vanilla' version is a bio-interface that allows us to access technology that gives us extra capacities, a bit like plugging an external hard drive into your laptop. You won't need to forget anything because you'll have petabytes of hard drive storage wired into your brain that you can recall data from at will. Difficult sum? Outsource it to your maths chip. In this scenario the consciousness is all biological, but it's got a load of new toys to play with.

The other flavour of mind-machine merge is where we unite human consciousness with machines, running part or all of our brains on this new platform. When I return to Boston, Ray Kurzweil will tell me we'll use technology to amplify and focus our intelligence and consciousness, replacing our brains with something more powerful, perhaps moving away from biology altogether. 'We won't be limited to a neo-cortex that fits into a less than one-cubic-foot skull, and we certainly won't run it on a chemical substrate that sends information at a few hundred feet per second, which is a million times slower than electronics,' says Ray. 'We can take those principles and re-engineer them and we're going to merge them with our own brains.'

This brings a whole new meaning to the phrase 'a meeting of minds.' The AInimals don't leave us behind or exterminate us, because we *are* the AInimals. 'We will become a merger between flesh and machines,' predicts Rodney Brooks. 'We will have the best that machineness has to offer, but we will also have our bioheritage to augment whatever level of machine technology we have so far developed. So we (the robot-people) will step ahead of them (the pure robots). We won't have to worry about them taking over.' In a 1995 paper entitled '*Will Robots Inherit the Earth?*' Marvin Minsky came to a similar conclusion: 'Yes, but they will be our children.'

All this makes many people feel uncomfortable. Most of us can make peace with some physical encroachment on our bodies. I'm very pleased with my contact lenses, for example. I'm glad Brooks's student can walk. And I think most of us are probably okay with brain interfaces that help the formerly disabled use new prosthetics. I'm truly heartened that Robin af Ekenstam can now eat a meal with a knife and fork using a robot hand controlled directly by his thoughts. Who would deny him the opportunity to regain his dexterity and sense of touch?

But merging our brains wholesale with technology? Or simply doing away with our neurons altogether and 'porting' ourselves over to silicon? I have to admit that writing this book would have been a whole lot easier if I'd had the Internet and the British Library wired into my head. And in a brain-meets-computer future you could read it as a download in milliseconds. Depending on how you view the world (or perhaps how busy you are), either the idea is strikingly brilliant or rather misses the point. It's the whole 'therapy vs. enhancement' argument I discovered at the beginning of my journey in Oxford.

It strikes me that one of the problems I have in making sense of the issues is that ideas as powerful as intelligent AInimals and man-machine mergers have the ability to instantly grab my imagination and emotions. I am instantly *there* somehow, imagining a dystopian or utopian future depending on my mood. But 'intelligence' and 'consciousness' are still mysteries, and although the fields of robotics, AI and neuroscience are doing much to unlock their secrets, I've seen nothing that leads me to expect that a robot army (or laundry detail) will be knocking on my door anytime soon. The journey to creating intelligent machines is progressing in baby steps, which means we should have time to negotiate our interactions with them as they slowly get cleverer (a bit like we do with our kids). This isn't to deny the extreme speed with which information technologies advance. Neither does it argue that we won't have the raw computational or manufacturing *power* to make such machines. But the ability of machines to understand *meaning* is moving much slower than the technology.

So intelligent robots look set to enter society rather slowly. And when they do, although a robot-human society will be *different*, I can't find a single shred of evidence that points to robots becoming our masters. Other human beings scare me far more than robots do. And if the robots do at some point start to vastly outstrip us intelligence-wise? We'll jump right on that bandwagon and go with them.

As my bus pulls out of Ithaca, George Church's parting words come back to me: 'Evolution is far from over. It's accelerating, and something is going to replace *Homo sapiens*. Is it going to be wetware, or is it going to be silicon? Is it going to be genetically engineered, or is it going to be electrically engineered? Or, is it going to be some hybrid?' My bet is on the hybrid, Rodney Brooks's merger of flesh and machines, which we're already seeing in medicine and will surely, slowly, continue. It sounds weird now, but the pace at which we'll accept machine encroachments on our bodies and minds (beyond those we already have) will probably come down to one simple principle (and our discomfort with it will melt quickly as a result) – and that is simply that the technology *doesn't suck.*

Not that long ago IVF was considered highly controversial. Indeed, the Catholic Church maintains its objection to the technique on the grounds that in IVF 'the generation of the human person is objectively deprived of its proper perfection: namely, that of being the result and fruit of a conjugal act.' However, mainstream opinion has generally got over its qualms about 'man playing God' when it comes to helping couples with fertility problems conceive. As George Church remarked, 'The instant it worked, and you started having people zipping about with their cute babies, everyone with reproductive issues was writing to their congressmen, and asking their doctor about it. They want that service, because they can see it, and see it can work. The argument flips a hundred and eighty degrees. It became *unethical* to *deny* access to IVF.' A brain link to Wikipedia sounds freakish now because it's not been done before.

It's important to note that no one will *force* you to implant anything. You can reject brain interfaces and night vision (and whatever else) just as you can decide to live without a mobile phone, television, the Internet and cheese. But it's likely that society, rather than rejecting these advances, will begin to insist it's *ethical* for us to have access to them, echoing the IVF story. You might take some enhancements and not others.

So for now the most intelligent machines on the planet are us. And it looks like in the future, they'll be 'us' too.

How quickly intelligent robots and man/machine interfaces become commonplace is linked to another technology that has the potential to radically reshape our world and our economics – a technology that even has the potential to spell the end of capitalism. It is perhaps the most powerful technology ever created, and will play a game-changing role in everything from

AI to medicine to manufacture. It's already beginning to enter our lives, but we're right at the beginning of its journey to ubiquity. Alternatively characterised as the harbinger of an unstoppable techno-rampage that could engulf us all, or a revolution that will reduce the cost of nearly everything, it promises to usher in a transformation in manufacturing more potent and far-reaching than the birth of the industrial age. It is the third pillar of what futurists call the 'GNR revolution' (the 'G' standing for 'genetics' and the 'R' representing 'robotics').

It is 'N' – nanotechnology. And it threatens to change the world in ways that will make your head spin.

CHAPTER 6

Invisibly small and magical

God dwells in the details. MIES VAN DER ROHE

'Most of them are higher dimensional beings. They don't *need* a spaceship to come here,' says the man with long grey hair sitting two tables away. 'You have to understand there *already is* a United Federation of Planets. We're being observed by twenty-six different alien races. Of course, the unethical ones – the Recticulans, the Greys, the Nordics – are ignoring the Prime Directive ...' There's a piece of Silicon Valley folklore that says if you really want to know what's going on in the technology industry, go and sit in a Hobee's restaurant, and listen to the conversation around you. I have clearly chosen the wrong seat.

But, as the conversation behind me drifts towards the frame-shifting potential of alien civilisations, I reflect that my own chosen topic, on this trip to California, is hardly less fantastical. Nanotechnology, at first glance, sounds much like science fiction. Yet it seems likely that it will radically reshape our future. At least, that's what future commentator and 'eco-pragmatist' Stewart Brand believes: 'The science is good, the engineering feasible, the paths of approach are many, the consequences are revolutionary-times-revolutionary, and the schedule is: in our lifetimes.'

Brand wrote that in 1991. Today nanotechnology is infiltrating nearly every sphere of human endeavour, from health care to construction. It has the potential to end industrial capitalism, revolutionise energy production, boost the power of medicine, deal a death-blow to nearly any resource crisis and ask you to review your relationship with your cleaning habits. And I'm in Silicon Valley to meet K. Eric Drexler, the man largely attributed with inventing it. In 1986 Drexler wrote a book called *Engines of Creation: The Coming Era of Nanotechnology* (now available for free on his website). And pretty much everyone who read it sat up and thought, 'Well, that changes everything.'

Nanotechnology has come to mean anything outside of biology that we try and do, engineering-wise, on the scale of nanometres – that is, on the scale of one billionth of a metre. It is what Drexler calls, 'the ridiculous "invisibly small and magical" definition.' And that's where I hit my first problem. I feel I need something a bit more concrete.

So I start on the maths. How small is a billionth of a metre? Think small. Very small. Go to the point where it's too small to comprehend and you're still *way too big*. Imagine a human hair. The average width of a hair is just over one-twelfth of a millimetre. You'd have to chop up the 30 cm ruler you had at school into 3,750 equal slices to get sections the same width as that hair. A nanometre is roughly eighty thousand times *smaller than that*, eighty thousand times smaller than the diameter of a human hair. To chop your ruler into nanometre-sized chunks, you'd have to slice it into *three hundred million* equal pieces.

It really is too small to conceive. Except a lot of people *are* thinking about things – indeed actively working on products – on this scale. But perhaps thinking only about scale is distracting me from what Drexler's original vision for nanotechnology is really about: manufacture and a bringing together of disciplines. As sometime nanotech historian Ashley Shew writes, nanotechnology 'is not one subject of study. Rather, it is a scale that has brought together researchers, innovators, and engineers from sometimes radically different fields. Disciplines collide on the tiny.' We're used to the collision in traditional manufacturing. Skills of design, robotics, materials science and electrical engineering come together to make everything from your car to your hairdryer. Those same disciplines and others are colliding again on an infinitely smaller playing field. This changes some of the rules of the game (how you make stuff) but the game (building the stuff) stays the same.

Focusing on *what can be made* as opposed to classifying things by their size is one reason the Industrial Revolution wasn't called the 'yard-long technology revolution' and it's the reason Eric Drexler now prefers the phrase 'molecular manufacturing' over the simpler 'nanotechnology,' which has become a catch-all for anything tiny. It's this nano-*manufacture* I've come to California to talk to him about, because – well, because it really could change everything.

For the promise of nanotechnology is based on the idea that if we can engineer things with atomic precision, we can develop materials and products with extraordinary properties. Changing the structure of something at the atomic level can cause it to exhibit unusual, often astounding properties. Nanotechnology is to matter what a phone booth is to Superman – it can bring about a transformation that unleashes superpowers.

Drexler opens his book *Engines of Creation* with this basic but powerful observation:

> Coal and diamonds, sand and computer chips, cancer and healthy tissue: throughout history, variations in the arrangement of atoms have distinguished the cheap from the cherished, the diseased from the healthy.

Carbon atoms don't just make coal and diamonds, either. Arranged one way they become the graphite in your pencil, and in another 'carbon nanotubes' – nanometre-sized tunnels with astonishing properties. At certain diameters, carbon nanotubes conduct electricity with incredible efficiency. Arrange their atoms slightly differently and you get super efficient 'semiconductors' (the key component in microchips) that generate far less heat and operate at much faster speeds than silicon-based counterparts. Another staggering property of carbon nanotubes is their strength – fifty times greater than steel – strong enough that a single strand the thickness of a hair could lift the weight of a family car.

By controlling the atomic structure of just one element, we create a wealth of opportunities: super light fuel-efficient aircraft, efficient transmission of power over huge distances, featherweight body armour, microchips running at speeds hard to imagine, even a cable that could stretch all the way into orbit providing a new route to space. And that's only carbon in its cape and bright red underpants. What might we achieve with other elements, both individually and in combination?

'The resulting abilities will be so powerful that, in a competitive world, failure to develop [them] would be equivalent to unilateral disarmament,' Drexler suggests. Guiding us through this change is, he says, 'the great task of our time.' He's not one to undersell his chosen area. Later he will tell me 'the reason I set out in the direction I did, and studied what I studied, is because I was convinced that civilisation is coming to a carrying capacity crunch that could create a lot of suffering. I could see a way to change that. So I set out to save the world.'

I am a little fortunate to get to meet Eric. In recent years he's adopted a policy of not meeting writers face to face, opting to answer questions by email. But he seems to like the direction I'm pursuing and, based on our emails, invites me to Silicon Valley – which is how I came to be sitting in this branch of Hobee's.

Part of Drexler's reticence to talk is no doubt due to the very rocky ride that he has been on since *Engines of Creation*. This includes a high-profile intellectual boxing match with a Nobel prize-winning chemist (Richard Smalley), and several run-ins with senior scientists who've questioned his credentials, suggesting 'he's an engineer who studied chemistry and thought about its possibilities, but perhaps did not take enough chemistry.'

Even his critics recognise his enormous contribution, although not all of them are taken with his Big Idea – whose first inklings came to him even before he wrote *Engines of Creation* but which he didn't fully articulate until his second book *Nanosystems* in 1992. It's an idea that still divides science, two decades on, with researchers often referred to as 'Drexlerian' or 'anti-Drexlerian' (terms Drexler himself loathes as personalised and anti-science). And the debate has extended beyond the science community, with Drexler's ideas twisted by the lens of headline-hunting journalism and appropriated for dramatic effect by science fiction writers. It's fair to say that over the years Drexler's Big Idea (and the man himself) have got under the skin of a lot of people. In 2004, *Wired* ran an article by Ed Regis (a largely pro-Drexler voice) with the strapline 'K. Eric Drexler was the godfather of nanotechnology. But the MIT prodigy who dreamed up molecular machines was shoved aside by big science – and now he's an industry outcast.' No wonder he doesn't like talking to writers.

Nanotechnology as an idea actually predates Drexler by a quarter of a century. Accepted history is that the field was born in a 1959 lecture entitled 'There's Plenty of Room at the Bottom' by the late great physicist Richard Feynman. For a long time the talk remained a little-quoted curio, but it has gained prominence retrospectively largely due to Drexler popularizing it.

During his talk, given to the annual meeting of the American Physical Society at the California Institute of Technology (Caltech), Feynman considered the possibility of writing the contents of the *Encyclopaedia Britannica* onto the head of a pin and concluded, if one could arrange atoms precisely to form letters, that not only could you get the encyclopaedia on the head of a pin, you could get all the information that mankind had ever recorded in books in a small pamphlet. Taking the idea further, Feynman pondered how much material would be needed to store all the words in all mankind's books if those words were converted to digital code, and deduced that, with a good deal of room to spare (atomically speaking) all of the information accumulated in all the books in the world could be held in a cube of material one two-hundredth of an inch wide 'which is the barest piece of dust that can be made out by the human eye.'

It still sounds incredible today. Imagine how it sounded in 1959. And indeed it was thirty years before we came to the next popular landmark in the nanotechnology story, when Donald Eigler became the first person in history to move and control single atoms in a repeatable and predictable way, arranging thirty-five xenon atoms on a bed of nickel to spell out the name of his employer: IBM. 'This capacity has allowed us to fabricate rudimentary structures of our own design, atom by atom,' wrote Eigler and colleague Erhard Schweizer. 'The possibilities for perhaps the ultimate in device miniaturisation are evident.'*

In 1959, the most obvious example of storing information on the nanoscale was DNA (the discovery of its structure had taken place six years earlier). 'That enormous amounts of information can be carried in an exceedingly small space is, of course, well known to the biologists,' Feynman reminded his audience. Your DNA comes in strands just two nanometres wide, curled up inside the cell nucleus – a structure that is about seventy-five times smaller across than Feynman's dustopedia.

Feynman's lecture reminds us that our own bodies are a walking collection of billions of nano-sized manufacturing plants churning out cell stuff. And if our cells can marshal materials on the nanoscale, then nanomanufacture is clearly possible. If it wasn't, we wouldn't be here. He went on to tell his audience that there was nothing in the laws of physics that forbade the manoeuvring of individual atoms. 'It is something, in principle, that can be done; but in practice, it has not been done because we are too big.'

Enter Eric, who is 'too big' but very smart.

Eric Drexler arrives at Hobees two minutes ahead of time, which gives me an excuse to move away from the *Star Trek* guy, who is now loudly holding forth about protecting against alien abduction by 'visualising yourself surrounded by gold energy.'

* In 2009, Eigler commented, 'It was more than a publicity stunt. Emotionally, for me, it was much more important. This is going to sound hokey, but it's the truth. IBM picked me up off the scrap heap of science and gave me every opportunity a scientist could hope for to be successful. As far as I was concerned, it was payback time.'

We find a seat by the window and order cinnamon tea. Taking his first sip, Drexler remarks, 'The tea is unusually strong right now,' before adding, 'it fluctuates.' This is Eric Drexler in a nutshell, exact in every detail, expressing himself by marshalling facts as he sees them first, and only rarely (and carefully) expressing opinion. He doesn't say the tea is 'good' or 'bad' but remarks on its strength. Or as he explains, before we begin our session: 'When I want to say that something is a good idea or a bad idea, desirable or not, that will be clearly flagged and said with qualification and trepidation.' He's a wiry fellow ('Down by about forty pounds from the nineties!' he tells me), with a well-kept grey beard covering a long thoughtful face, in which two piercing brown eyes seem to say, 'Yes? And what more?'

Eric Drexler is not one to be satisfied with half a story, or abstract musings. He tells me he gets annoyed 'when scientists suggest their work implies new insights but don't *explicitly state* what those insights are.' You'd better know the history of your profession too. So, if you work for NASA and invite Eric for lunch but don't recognise the name 'Konstantin Tsiolkovsky,' he'll be mildly disappointed in you (as he tells me he was just a few weeks earlier).

I hadn't heard of Tsiolkovsky either (I don't work for NASA so I'm let off), but the reclusive Russian schoolteacher who was writing about multistage rockets for reaching Earth orbit as early as 1903 becomes a recurring theme in our conversation. Drexler sees little excuse for being ill-informed, educating himself on a wide range of topics (his blog contains a post entitled 'How to Understand Everything – and Why'). In this sense he is mildly terrifying. But in person he's neither prickly nor intimidating. He's certainly exact, but he's generous too – and interested. He wants to know all about my trip to see 'the remarkable George Church' and is interested too in my investigations into robotics at MIT and Cornell. (His opinion is that a human-level Artificial Intelligence is 'inevitable.') Maybe his chatty and curious mood is because he's had a bit of a renaissance – Eric's Big Idea is back on the table.

This idea is the 'nanofactory' – a desktop-sized programmable device that, in principle, could manufacture, with atomic-level precision and absolute fidelity, any number of hugely powerful and useful things using a cheap supply of simple 'chemical feedstocks.' In the introduction to the twentieth anniversary edition of *Engines of Creation*, he writes that

> A 10-kilogram factory will be able to produce 10 kilograms of products in hours or less – a stack of billion-processor laptops, a package containing a trillion cell-sized medical devices, or a roll containing hundreds of square meters of tough, flexible stuff that converts sunlight to electric power. It seems that raw materials will be the main cost of production. At a dollar per kilogram (a typical price for industrial feedstocks today) the solar-electric material would cost about one cent per square meter, and the computers would cost about a dime.

By approaching matter in the way computers treat information, we move from a world where factories specialise in making one product to *programmable* factories, or 'matter printers' that can make myriad products as long as they're fed the right design and raw materials – a multi-purpose machine creating products to order at a fraction of the cost of today's methods. Just as computers 'form complex patterns of the elementary building blocks of information' writes Eric, molecular manufacturing systems will 'form complex patterns of the elementary building blocks of matter.' Manufacturing becomes an *information technology*, a world where we can email product specifications to each other, and our desktop 'matter printers' make them.

'So, you're taking the Industrial Revolution model of the factory, putting it on my desktop, and I can program it to make, almost anything? It's Industrial Revolution Two-Point-Oh but computable?'

Eric thinks for a moment. 'Yes. I think the comparison to the Industrial Revolution is a highly appropriate framing, not only because of the factory model, but also because when you change how things are made, and therefore what *can* be made, this has profound effects on everything.'

One of those 'profound effects' could be the end of capitalism. I'd thought I'd left the *Star Trek* comparisons behind when I moved tables but if Eric's right then many of the ideas Gene

Roddenberry explored in his iconic sci-fi series could come to pass. The argument goes something like this: How much we pay for things is based on two key factors, what they cost to make and how scarce they are. Nanotechnology has the potential to deal a sucker punch to both.

With Eric's nanofactory, raw materials are cheap chemical 'feedstocks.' Common elements like hydrogen, carbon, nitrogen, oxygen, aluminium and silicon seem best for constructing the bulk of most structures – vehicles, computers, clothes and so forth – as they are light and form strong bonds. And as Eric asserts, 'because dirt and air contain these elements in abundance, raw materials can be dirt cheap.'

In his future, products are assembled by nanofactories following programs, not by human physical labour, so there's no need for wages. Need more capacity? Because the nanofactory is programmable it can quickly build a copy of itself (at a tiny cost). Among the other products it can make, of course, are devices for generating energy, like super-efficient solar cells – so the cost of energy becomes negligible. It's a world where nearly everything is cheap to make out of abundant raw materials. And if we really do need some esoteric scarce element we can't manufacture an equivalent for? Eric's answer is cheap nanomanufactured spacecraft that mine the solar system. The asteroid belt has resources of rare metals that make Earth's complement look like half a potato brought to the harvest festival.

'It's pretty fantastical,' I say. 'Is it possible?'

Eric gives one of his exact and heavily qualified answers (my translation in brackets): 'I'm looking at phenomena we know exist in the world [he means the laws of physics, chemistry and biology] ... we don't yet know how all these work but that doesn't necessarily matter [engineers don't have to know everything about the chemistry of materials to build bridges]. If you apply large design margins you get a very reliable picture of what you can achieve if you actually had the resources – technology, money, people to put it together. It's a bit like Tsiolkovsky's rockets, though – this might be rather far down the line [you need a *very big* research budget and a lot of patience].'

Of all the things I've investigated so far, nanotechnology seems the closest to the realm of science fiction. And there is actually a sub-genre of sci-fi called 'nanopunk' which riffs on future worlds changed by nano-engineering. José López from the University of Ottawa makes the point that sci-fi and nanotechnology are deeply entwined, due to the field's 'radical future orientation' which he argues 'opens up a gap between what is … possible today and its inflated promises for the future.' It's a view I'm surprised to hear Eric partially agrees with.

'If anyone were trying to build a nanofactory today, I would say I have no idea what they think they are doing. It would make no sense. It'd be like trying to pick up rocks on the moon when we haven't yet put a satellite in orbit.' He won't say when it's going to happen, just that it probably will.

On the way there, lots of things will change. Already, 'nanoproducts' are coming to market which, if they live up to the hype, could radically reshape whole industries. Even if these first-generation products turn out to be marketing over delivery, the underlying science is sound, suggesting it's only a matter of time before they hit pay dirt. The first major casualty of the Nano Revolution?

Cleaning products.

We're used to thinking of cells as gloopy, semi-liquid things that alter their shape as they bustle around each other, but not all cells are like that. Diatoms, a single-celled form of algae, have *hard* cell walls made of silica (everyday forms of silica include sand and quartz). When diatoms die, they leave behind tiny glass skeletons staggering in their intricacy and beauty. These skeletons come in two parts, one half fitting snugly over the top of the other, like the two parts of a Petri dish that children use to collect bugs. In diatoms, the sides of these overlapping parts contain tiny nano-scale holes, and by moving these two parts in relation to each other, a diatom can control how big an aperture it has on the outside world, allowing things it wants in (for example, molecules of oxygen) but keeping out things it doesn't (like bacteria). It's the tough, glass nanomembrane that keeps the diatom safe.

In recent years there have been numerous reports of manufactured silica coatings that mimic the diatom's defence mechanism. Like a diatom's cell wall, these coatings contain tiny pores, nanometres across, that keep bacteria, dirt and viruses at bay, while at the same time letting the material underneath breathe. These coatings themselves are just one hundred nanometres thick, making them invisible yet flexible enough to cover anything from tanks to tank tops. Potential applications for this 'liquid glass' include everything from battling infection in hospitals, to keeping monuments pristine, to protecting against rust, to making sure your clothes never get dirty. To clean anything you'll just need water. Goodbye most detergents and cleaning products (and a whole industry with them).

This is just one example of hundreds. Suddenly I'm seeing the 'hand of nano' everywhere.

A team at the University of California, for instance, is working on a technique that dispatches antibodies to sniff out cancer cells and then marks them with a gold 'nanosphere.' When exposed to a particular frequency of infrared light, the nanospheres heat up and literally bake the cancer cells to death. Following an injection of the antibodies 'you could send a person home, have them shine a laser on the specific part of the body with cancer for a couple of weeks, and they could be cured,' says professor Jin Zhang. Eric Hoek and colleagues at the California NanoSystems Institute are working on new nanomembranes that could reduce the cost of desalinating water, solving the problem of predicted fresh water shortages. And researchers at Stellenbosch University's Water Institute in South Africa have put carbon and nanofibres into common tea bags to create a bottle-based, impossibly cheap water filtration system that removes pathogens from dirty water sources – a potential boon for millions of Africans.

Nanomaterials are also about to see widespread use in the construction industry to 'enable novel applications ranging from structural strength enhancement and energy conservation to antimicrobial properties and self-cleaning surfaces' according to a report from Rice University and the University of California. In Greece, the Intelligent, Safe and Smart Built (ISSB) project is developing a house for earthquake zones that uses nanomaterials to self-heal cracks: nano-polymer particles that become liquid when squeezed under pressure, flow into the cracks and then harden into a solid material.

I could go on. The Project on Emerging Nanotechnologies (a partnership between the Woodrow Wilson International Center for Scholars and the Pew Charitable Trusts) keeps a watching brief on nanotechnology-based consumer products. The list is already over a thousand products long and growing every day. It includes computer memory and microprocessors, numerous cleaning products, bandages with nanotech 'antimicrobial' barriers, vehicles (from cars to jet-skis to baby carriers) and sports equipment with lightweight but strong body parts, anti-odour socks, toothpaste, air filters, sunscreen, soft but puncture-resistant tyres, 'antibacterial' medical tools and kitchenware, fabric softeners, pregnancy tests, cosmetics, stain resistant clothing and pet furniture, long-wearing paint, bed-ware, guitar strings that stay sounding fresh thanks to a nano-coating, white goods, a 'soil wetting agent' that allows previously infertile soils to retain water ... and (it seems to me) a disproportionate number of hair straightening devices.

Another promising nanotechnology is 'DNA origami,' using DNA as *raw material* for making nanostructures (i.e., being used as a *structural substance* and not as a store of information). DNA makes a good building material because we know how it snaps together (A's love to bind to T's, and the G's to C's). Paul Rothemund of Caltech used this two-way binding as a way to rivet bits of DNA together in interesting ways. One of his first demonstrations of this technique was to create a map of the Americas one 200-trillionth of their actual size.

But there are more useful applications than 'DNA nano-art.' One example is building tiny nanoboxes with lids and 'molecular locks' that might be used to contain drugs until they can be delivered to just the right cell – and then unleashed with devastating accuracy. In another example, Rothemund and colleagues used DNA 'staples' to carry some of those wondrous carbon nanotubes into position to form a tiny circuit switch even smaller than those found in modern microprocessors.

In May 2010, chemists at New York University and China's Nanjing University used this DNA origami technique to create a nano-sized factory shop floor with an assembly line that could be followed by a 'DNA walker' that picks up nanocomponents as it moves along the line. By the time the walker gets to the end, it has been added to, just like a car chassis slowly turns into a car as it makes its journey from one side of an automobile factory to the other.

The food industry, too, is beavering away on nanoresearch, looking into new molecular tools for detecting and treating crop diseases and increasing the ability of plants to absorb nutrients. 'In the near future nano-structured catalysts will be available which will increase the efficiency of pesticides and herbicides, allowing lower doses to be used,' concluded the 2006 Nanotechnology in Agriculture and Food report by the EC-funded Nanoforum. In the same report, the idea of interactive food is discussed, a product that will 'allow consumers to modify food depending on their own nutritional needs or tastes. The concept is that thousands of nanocapsules containing flavour or colour enhancers, or added nutritional elements (such as vitamins), would remain dormant in the food and only be released when triggered by the consumer.' Another idea is to create healthy versions of traditionally unhealthy treats – donuts, pretzels and cheesecakes that taste as bad as you like but contain nice low levels of fat, salt and sugar.

So there's *already* a nanotechnology revolution going on, even if none of it looks, right now, like Drexler's desktop revolution. And it's growing apace, as even a cautious commentator like Professor Richard Jones, a one-time senior nanotechnology adviser to the UK government, and author of *Soft Machines: Nanotechnology and Life*, acknowledges:

> The main impact of 'nanotech' ... will be the reduction in cost of most manufactured products. This is already happening

> with electronic goodies (TVs, PCs, music systems, etc.). This same progression is likely to happen in biomedical and normal manufactured products as well (cars, houses, etc.).

In short, nanotechnology is the future of manufacture. Unless, that is, we all get eaten by Grey Goo.

Grey Goo? Seriously. You can't discuss nanotechnology without someone mentioning this particular version of Doomsday soon after. For in *Engines of Creation*, Drexler imagines tiny atomically engineered nanodevices called 'replicators' that carry a program which instructs them to continually make more of themselves:

> the first replicator assembles a copy in one thousand seconds, the two replicators then build two more in the next thousand seconds, the four build another four, and the eight build another eight. At the end of ten hours, there are not thirty-six new replicators, but over 68 billion. In less than a day, they would weigh a ton; in less than two days, they would outweigh the Earth; in another four hours, they would exceed the mass of the Sun and all the planets combined.

The popular analysis of this was that we will be overtaken, eaten up by trillions of tiny mechanical robots – neither necessarily grey or gooey, but the name stuck. And indeed, *Engines of Creation* is replete with sombre warnings ('Dangerous replicators could easily be too tough, small, and rapidly spreading to stop ... we have trouble enough controlling viruses and fruit flies.') Drexler suggests that a wait-and-see policy could 'cost many millions of lives, and perhaps end life on Earth,' rendering our future, 'exciting and short.'

But his motivation for raising the Grey Goo issue was, he tells me, simple 'civic duty' – a warning for us to be vigilant in case someone attempted to deliberately manufacture something Grey Goo-like. Meeting Eric gives me the strong impression there is nothing sensationalist about him. His catalogue of the potential benefits and downsides of nanotechnology is simply that – a catalogue with very little emotional weight attached to any entry. These are statements as controversial or emotional to him as 'the tea is unusually strong.' Other people (understandably) didn't take it that way. In a bruising public exchange of letters with Drexler, Richard Smalley, a Nobel prize-winning chemist famous for his work on those carbon nanotubes wrote:

> You and people around you have scared our children. I don't expect you to stop, but I hope others in the chemical community will join with me in turning on the light, and showing our children that, while our future in the real world will be challenging and there are real risks, there will be no such monster as the self-replicating mechanical nanobot of your dreams.

I ask Eric why the Grey Goo idea became so prevalent. 'Two reasons,' he responds. 'The first is because I'd used a biological analogy and in biology you do have small things that build other things like themselves. Grey Goo makes intuitive sense, even though it doesn't make scientific sense.' The reason being that all self-replication (including that performed by cells) requires some method of collecting raw materials. If you stopped eating and breathing your cells would soon cease dividing. So for any nano-machine to self-replicate it would need to have the ability not only to continuously find raw materials, but also free them from whatever they're already part of. This is actually spelt out even within the pages of *Engines of Creation*: the scenario in which replicators 'exceed the mass of the Sun and all the planets combined' has them 'floating in a bottle of chemicals' and their dominance would only be feasible 'if the bottle of chemicals hadn't run dry long before.'

The second reason? 'Grey Goo makes a good scare story, as long as you leave out the science.' Indeed Michael Crichton's 2002 novel *Prey* is just this scare story, radically massaging the science in the service of drama. 'But if you want a system that manufactures things look at an automobile plant,' says Eric. 'Can it make another automobile plant? Not remotely. And an automobile can't make an automobile. Is there, in fact, any self-replicating system in a box? Has anyone set out to do it?

No. Because there's no motivation.' It doesn't work as a model of manufacture because it makes the whole thing staggeringly inefficient.'

Then Eric puts one of the strangest images of my trip into my head. 'It's hard to design a car that gets pregnant and replicates itself,' he says. 'That's a difficult engineering challenge. It would be extraordinarily difficult to make, there's no reason to and the result would be very inefficient.' The world isn't built to support unstoppable self-replicating machines (on any scale) any more than it's built to support an evil strand of amoebas that could spiral out of control and eat up the rest of life on Earth.

Looking back at Grey Goo now, he says the idea is 'stupid' and I get the impression he rather wishes he'd never mentioned it. Ironically Grey Goo's most dangerous act of self-replication has been as an idea – and certainly one that contributed to Eric's spell in the cold.

The field split as former enthusiasts began turning against Drexler. Richard Smalley (who accused Drexler of scaring children in 2003) admitted that reading *Engines of Creation* 'was the trigger event that started my own journey in nanotechnology ... [but] after a while I thought I saw what might be some problems. The more I thought about it, the more troublesome they appeared. Finally I ended up thinking, it's just hopeless.'

At the root of the anti-Drexlerian position is what chemists know about how things fit together at the molecular level. They point out that when you get down to the nano-scale there's a mêlée of natural chemical reactions with molecules and atoms attracting and repelling each other and out of control physical collisions (for nano-components a rogue piece of dust is like an iceberg thousands of times bigger than that which sank the Titanic). While biology has certainly developed nano-manufacture, most notably in the form of cells, it's had to do it within the bounds of what chemistry will allow. Try to go too far against that and chemistry will be knocking on your door.

Every atom interacts with every other, including those that make up Drexler's nanofactory. For the chemist critics the nano-scale looks far too much like the mosh-pit at the front of a death-metal gig for it ever to be a place to build an assembly line.

Eric counters, 'People say you're not thinking about the fact these molecular things stick together. Well that would be like saying someone who's designing aircraft isn't thinking about gravity. It's a fundamental part of the problem.' This is the premise of *Nanosystems* where he addresses a good deal of that 'messy chemistry'. And this, too, is a book that has major critics. Julius Rebek, a professor at the Scripps Research Institute in San Diego and by anybody's reckoning a serious chemist, has commented that sections of *Nanosystems* are 'not science' but 'show business.' This wasn't the first time Rebek had criticised Drexler's scientific credentials. Of Eric's original MIT doctoral work he remarked, 'it showed utter contempt for chemistry. And the mechanosynthesis stuff I saw in that thesis might as well have been written by somebody on controlled substances.' Ouch.

A more measured voice is Professor Richard Jones, mentioned earlier, who says *Nanosystems* is 'a carefully written book', that 'Drexler's proposals for radical nanotechnology do not obviously break physical laws' but that 'many proposals in *Nanosystems* are not fully worked out, and many vital components and mechanisms remain at the level of "black boxes."' Back in Hobee's restaurant Eric is agreeing. 'What I'm saying is: this is the current understanding of how one might organise a molecular manufacturing system. And yes it's still in outline. We've got some engineering to do ... Note that I dedicated *Nanosystems* "To experimentalists, engineers, and software builders" – they do the hard parts.'

It is a familiar debate, science vs engineering. Scientists tend to want to know everything. Engineers want to know enough to get building. As Eric said, it's not about 'are there some things that won't work?' but 'are there enough things that will?'

However, after a long time in the cold, Drexler's nanofactory idea was about to have a partial renaissance. In 2006 a joint report from the National Academy of Sciences, The National Academy of Engineering and The National Research Council concluded that 'manufacturing processes with some capability to pattern structures with atomic precision' might one day be possible and research into programmable manufacture at the molecular level should be cautiously encouraged.

That report's final paragraphs are carefully worded, pointing out that the future performance of many theoretical components of molecular manufacturing cannot be 'reliably predicted at this time.' Nonetheless they're clear that the theoretical feasibility of a Drexler-like system cannot be argued against. To make the point they reference the now familiar Russian schoolteacher and rocket fan. 'This work is currently outside the mainstream of both conventional science (designed to seek new knowledge) and conventional engineering (usually concerned with the design of things that can be built more or less immediately). Rather, it may be in the tradition of visionary engineering analysis exemplified by Konstantin Tsiolkovsky's 1903 publication, *The Exploration of Cosmic Space by Means of Reaction Devices*.'

Elsewhere, hints that programmable nano-manufacture may one day be possible began to surface. Nearly twenty years after Don Eigler announced he'd made an IBM logo out of xenon atoms, researchers at the same lab (along with the University of Regensburg in Germany) worked out the forces needed to manipulate specific atoms on specific surfaces. 'This result provides fundamental information about atomic scale fabrication and could pave the way for new data storage and memory devices,' said IBM researcher Andreas Heinrich. 'Our mission is to create the foundation for what could someday be called the IBM nanoconstruction company.'

Let's assume for the moment that Eric's ambition is possible – that at some point in the future the nanofactory is coming.

The implications are enormous. Humans will no longer have to make stuff. Manufacture will be a design-only profession. What you can imagine (within the laws of physics) will be makeable, including machines that generate almost free, clean energy, and computers with calculational powers that will appear almost magical, millions of times faster and millions of times more energy efficient than their counterparts today. Manufacturing capacity will be, essentially, unlimited. New products will be able to be prototyped within days rather than years. Ideas can become matter. We will want for nothing.

Then again, there is the end of the world to think about. There are at least five versions of a nanotech-made apocalypse.

(1) *Economic meltdown.* One nation gets the nanofactory ahead of everyone else. Suddenly they don't need to buy very much any more. So, for instance, the US stops buying goods from abroad. Economies dependent on American trade go into a tailspin, igniting a global economic crisis.

(2) *Nanotechnology arms race.* Nations terrified of the threat posed by foreign powers who might possess this technology ahead of them race to produce their own version, driven by the same desires that saw nuclear arms proliferation. Once again, Mutually Assured Destruction via powerful nano-engineered weapons is on the table and the world lives in fear of, or worse, suffers an all-out nano-war.

(3) *Welcome back Hitler.* At the 2008 Global Catastrophic Risks conference, Mike Treder, co-founder of the Center for Responsible Nanotechnology, said, 'If you can make any product you want on your desk in your home, what happens to all the people whose jobs rely on finding raw materials, manufacturing those products, delivering them – transportation, storage, wholesale, retail – all of those people might someday be left jobless ... Such social disruption can leave a vacuum that allows a powerful charismatic personality to take hold. We've seen that happen before ... and it could lead to a situation of global tyranny.'

(4) *The democratisation of violence.* Mike Treder again: '[It can lead to] smaller groups or even individuals [wielding] greater

destructive power. If any technology was ever primed to do that, this is one.' Ideological nutters have the ability to kill a lot of people because their nanofactory can manufacture weapons of unimaginable destruction.

(5) Police state. Nanotechnology allows you to be watched without your knowing. A billion unseen cameras, microphones and tracking devices could be used to observe and then suppress a populace. This is Eric's personal 'favourite.' 'That's what people should be scared of,' he tells me. 'Not terrorism but the ability to *suppress anything like terrorism.*'

And of course, these could all combine in some unimaginable way. Nanofactory-inspired global economic meltdown and social disruption lead to the rise of dictators who inherit the fruits of a nanotech arms race and fine-grained surveillance. We settle down into a world of fear and constant monitoring.

Oh dear.

The problem with all these scenarios is they're naturally built on our understanding of history. We're using a lens shaped by the past to look at the future. But with a nanofactory revolution, the lens is just wrong. It can't help us see a world without scarcity because we've never had one. If we learn to control matter in the way Eric suggests, then we're making a shift as fundamental as our primate ancestors did when the idea of using tools first occurred to them. As primitive as using a jawbone for a club looks to us now, so industrial manufacture will look to our descendents (or maybe us, if it comes soon enough). Any nano apocalypse we imagine using our old lens is equally flawed.

Two things seem clear. First, the Nano Revolution won't be owned by any one nation. Nanotechnology is already an international discipline. The examples in these pages come from Asia, the United States, Africa and Europe. There are nanotech initiatives and hubs in the Middle East, India and South America. There is no dominant party. Second, it won't happen overnight. As I write, it's nearly a quarter of a century since Drexler laid out his vision in *Engines of Creation* and Eric himself admits anybody trying to build a nanofactory would be entirely misguided. Just like Konstantin Tsiolkovsky and his rockets in 1903, we're still a long way off having the engineering to get us from here to there. Nanotechnology will enter our lives in increments. The world will not turn on a penny.

Taking these two observations together it is hard to imagine either a nanotech-inspired economic meltdown or a scenario where one nation suddenly acquires the nanofactory and then either suppresses all knowledge (worldwide) related to nanofactory development, or uses the technology to suppress/invade other territories quickly enough to ensure some kind of draconian order. The other reason a war brought on by nanotech is hard to imagine is answered when you think about the question 'Why do we fight in the first place?'

The most convincing theory I've come across is related to the Earth's 'carrying capacity' – as Eric himself mentioned early in our conversation. Human beings have a habit of outstripping their natural environment, and when they do, they start to compete for what have become limited resources. We want more of the land, food, minerals, labour or fuel, and suddenly there isn't enough. Nations try to ring-fence sufficient resources and sometimes that means taking them from someone else. As the population rises we have two choices: find a way to increase carrying capacity; or fight (justifying the latter with some form of ideology).

But what has this to do with nanotechnology? The same Stewart Brand who suggested the impacts of nanotechnology would be 'revolutionary-times-revolutionary' provides a link in his book *Whole Earth Discipline: An Ecopragmatist Manifesto*:

> Peace can break out, though, when carrying capacity is pushed up suddenly, as with the invention of agriculture, or newly effective bureaucracy, or remote trade or technological breakthroughs.

Nanotechnology can potentially (and dramatically) increase carrying capacity and, crucially, the distribution of resources –

because everyone has the raw materials needed to make whatever they need, including food.

And, as Eric maintains, at the same speed at which nanotechnology offers us more inventive weapons, it's offering a reason not to use them: 'Scenarios where the motivation is to win resources are incoherent.' This makes Nanotech Apocalypse 3 (*Welcome back, Hitler*) hard to imagine. Dictators require some inequalities upon which to sell their ideologies, but technology tends to reduce inequality. In fact, it's equally valid to argue that the greatest risk to society is not in *getting* the nanofactory, but in not getting it *soon enough*.

Reflecting on the two other concerns – ubiquitous surveillance, or putting extreme weapon-making power into the hands of idiots – I'm transported back to Harvard with George Church talking about the power of synthetic biology and the need for a scientist-led movement for oversight and licensing. Nanotechnology is a younger science and is only just waking up to the need to work out a regulatory and ethical framework that brings us the benefits, but avoids the pitfalls. Regulation tends to get off the ground after we've had a few accidents and outrages; and nanotechnology, like synthetic biology, will have its fair share.

The 'good' news is that those first accidents will be as much a reflection of the technology's advancement as its successes. There is no chance of your crazy flatmate getting hold of a nanofactory and vaporizing the neighbourhood with a new nanoweapon anytime soon.

As for ubiquitous nanosurveillance, there is a black hole in legislation that also needs to be addressed as the technology matures. That said, surveillance and anti-surveillance technologies will likely develop in step. Today you can buy cheap anti-spy camera scanners that let you know if there's a device transmitting wireless camera signals in your vicinity and find out where it is (either by homing in on its signal or using LEDs to reveal the glint of hidden camera lenses).

I suspect therefore that nanotechnology will follow synthetic biology's example. As the science becomes more 'real,' regulation will evolve, and some people will die because we didn't work out all the risks in advance.

Does Eric have an idea about when we should *start* worrying about the legislation? How far off is the nanofactory revolution? 'I'm not at all inclined to talk about time frames,' he says. 'What I will say, is that the scientific knowledge and the tools are in place such that with a lot of creativity we can get beyond *some* of the difficulties of designing and using molecules in the near future.' In other words, don't hold your breath.

But in some ways it doesn't matter if he is right or wrong. In a blog post titled 'Even if Drexler is wrong, nanotechnology will have far-reaching impacts,' Richard Jones asserts:

> Followers of Drexler are in danger of finding themselves in denial about the potential impact of ordinary, evolutionary nanotechnology, because of their devotion to their brand of nanotechnology's one true path . . . It would be ironic if, in thirty years, the Drexlerites find themselves still waiting for a revolution that's already happened.

He is surely right. Even if we don't get as far as the nanofactory, nanotechnology will continue to be exploited in energy and food production and in creating new materials. The nanotech revolution has started. And as Jones goes on to observe, 'Perhaps we should thank Drexler for alerting us to the general possibilities of nanotechnology, while recognizing that the trajectories of new technologies rarely run smoothly along the paths foreseen by their pioneers.'

Drexler, in my book, will be remembered as a dreamer, indeed a visionary: the man who got the nanotechnology ball rolling. And that alone may well realise his ambition to 'save the world.' In the near future, nanotechnology looks set to underpin revolutions in energy production, medicine and sanitation. It has a very real potential to make our world cheaper, healthier, more sustainable and a lot cleaner.

CHAPTER 7

The biggest, baddest kid on the block

After one look at this planet any visitor from outer space would say 'I want to see the manager'. WILLIAM S. BURROUGHS

I'm on an eight-hour train ride that will take me deep into the California desert. As the scenery rolls past, I can feel my mind trying to stretch to accommodate the idea of societies remade by nanotechnology. Once more, I find my long-held views about what the world is and how it could be are being challenged. It's invigorating but tiring. I'm less than halfway through my journey but already I feel like I need a holiday.

But the mind-bending is far from over. My next stop will challenge even the idea of what a holiday can be ... I'm going off-planet. Which is what brings me to a desert town and the lounge of a roadside bar. Ozzy Osbourne is playing loudly over the jukebox, too loud for me to concentrate on my book. There are no windows. Finishing my dinner, I find a seat at the bar in the saloon. Ozzy isn't quite so obtrusive here. There are two other, silent, customers. I order a beer and start reading.

'Never seen anybody with a book in here before,' says the bigger of the two men. I'm beginning to get the feeling that this isn't the kind of bar that holds weekly philosophical meetings.

'I don't have to read,' I say, turning to him. 'Let's talk.'

Travelling alone gives you a hunger for conversation.

'You live here?' I ask.

'In the bar?'

'No, Mojave. The town.'

'Sure,' he says and falls silent, mulling over his beer.

'Were you born here?' I ask, persevering.

'No, I just moved from Los Angeles.'

'That's a hell of a change.'

Mojave, it has to be said, is hardly a thriving metropolis. It's essentially one road with a McDonald's, a Carl's Junior, a Burger King, a Denny's, a gas station and a clutch of motels. As for the social scene, I'm looking at it.

'Why did you move?' I ask.

'Less trouble here,' he says resolutely.

'You had trouble in Los Angeles?'

'I was in jail.'

Now I understand. I'm in a road movie from the seventies. Any moment now Clint Eastwood is going to walk in and there will be a gunfight. I'll dive behind the bar as bullets shatter the bottles above me.

'Can I ask what you were in jail for?' I inquire, trying to sound nonplussed.

On the subject of violence he becomes a bit more vocal.

'Attempted murder. I bashed someone over the head during a robbery.'

To my knowledge this is the first person I've met who's been to jail for trying to kill someone. Clint is late. Where is Clint?

'I needed money for drugs,' he explains, warming to his theme. 'I don't do drugs any more. There's a lot less drugs in Mojave.'

To be honest, there's a lot less *everything* in Mojave, with one notable exception. Just over a hundred years before my journey to this town in the heart of the California desert, two brothers chose an equally remote location – Kitty Hawk, North Carolina (nearly three thousand miles east of here) – to test out their new machines. Relative isolation freed them from the prying eyes of reporters and competitors. They were neither wealthy nor government funded, but what they did changed the world. The brothers in question were bicycle engineers Wilbur and Orville Wright who, late in 1903, achieved controlled, powered flight. I've come to Mojave to meet some of their modern-day equivalents.

On April 12, 1961, Russian cosmonaut Yuri Alekseyevich Gagarin became the first man to orbit our planet, his achievement part of a geopolitical 'superpower' space race that has dominated the tone of human spaceflight ever since. Human-carrying space missions are a matter of national pride and ego, open only to a select few. In the nearly fifty years since Gagarin's flight, less than 550 people have had the opportunity to follow his example.

In 1968, the film *2001: A Space Odyssey* imagined manned missions to Jupiter, hotels in orbit and a spaceflight industry that looked like an extension of the aviation business. It seemed like a fair prediction at the time. The same year, as Frank Borman, James Lovell and William Anders orbited the moon in Apollo 8, Pan American airlines started taking the names of passengers for commercial lunar flights. Why not? After all, it took just thirty-six years from the Wright brothers' breakthrough to Pan Am offering a transatlantic passenger service between New York and Marseilles. But Pan Am went bankrupt in 1991, leaving a reported 93,000 names on its lunar reservations list.

Paradoxically, the roots of our failure to open up space are the same roots that led to our early successes in travelling over the Kármán line (the point at which the air gets too thin for normal aircraft to fly – and you need a *space*craft to keep going). The space race wasn't driven by a desire on the part of Russia or America to usher in new technological, commercial or humanitarian possibilities. It was a militaristic bun fight.

'Apollo was never designed to open up space – it was designed to make us look like the biggest, baddest kid on the block,' says

Rick Tumlinson, co-founder of the Space Frontier Foundation, which believes it is possible to achieve 'large-scale industrialisation and settlement of the inner solar system within one or two generations,' an ambition they believe has been stifled by a 'centrally planned and exclusive US government space program.'

For Tumlinson, our failure to grant widespread access to the solar system is due to politics, not possibilities. Ever since the Apollo missions, NASA has dominated the debate and popular understanding of space, reinforcing the notion that human spaceflight is so costly and risky it is only achievable with huge levels of public funding. By its own estimates the average cost of launching the Space Shuttle 'is about $450 million per mission.'

Space exploration became a poster boy for vainglorious excess in a troubled world. Not that there haven't been tangible benefits for humanity. Our exploitation of the stars has led to better understanding of our climate systems. It has helped reduce our consumption of fossil fuels thanks to the existence of GPS routing; bolstered global communications networks; and the development of new drugs and materials have stemmed from experiments in low gravity. But for most people, and most governments, the price tag and the rewards have a less than direct correlation.

But what if access to space became radically cheaper and started to deliver real value-for-money in the eyes of the world? For instance, the moon is an abundant source of Helium-3, a non-radioactive form of helium that has been proposed as a clean nuclear fuel. The asteroid belt is awash with precious minerals. What if we could put an end to many resource shortages by mining the lifeless rocks of the solar system? In 2001, NASA's Near-Earth Asteroid Rendezvous (NEAR) mission landed on the asteroid Eros. Analysis of this twenty-one mile long, eight-mile wide and eight-mile thick hunk of rock revealed it contained 'precious metals worth at least $20,000 billion.' (Of course if all those minerals ever made it to Earth, they would flood the markets, so the twenty-trillion-dollar valuation would plummet – but there's still potentially big money in space-mining.)

Others suggest that if we could afford to get more of our world leaders to see the Earth from space, they might embrace environmentalism and see the value of increased cooperation among nations. Apollo 11 astronaut Mike Collins said, 'The overriding sensation I got looking at the Earth was, my God that little thing is so fragile out there,' while Apollo 8's Frank Borman told *Newsweek* that looking back at the Earth brought home the thought that 'this really is one world and why the hell can't we learn to live together like decent people?'

For others, space is about survival. Within a few billion years the Earth will be consumed when the sun goes supernova. But long before that, there is the probability of a cataclysm befalling us, such as the asteroid that wiped out the dinosaurs. Science fiction author Larry Niven once said, 'The dinosaurs became extinct because they didn't have a space program. And if we become extinct because we don't have a space program, it'll serve us right!' (In June 2010 NASA announced that its Kepler space telescope had found candidates for seven hundred and six Earth-like planets after surveying one hundred and fifty-six thousand stars in a single patch of sky.)

But whatever your view on the possible reasons for going to space – whether it's to find resources or inspire environmental awareness – the sticking point is cost. I've come to Mojave to meet the entrepreneurs and engineers who are hoping to solve that problem.

The sign says 'Mojave Air and Space Port'. But if that conjures up a futuristic image of gleaming towers, think again. The spaceport is a sprawling industrial park populated by nondescript sheds and factory units. As airports go, it's on the shabby end.

Still, it was here that Richard Branson's Virgin Galactic grabbed headlines in 2004 by launching *SpaceShipOne*, winning the $10 million Ansari X Prize for Spaceflight, offered to the first commercial vessel able to take three people over the Kármán line and return them to Earth twice in fourteen days. The total cost of this *entire development* was rumoured at between twenty and thirty million dollars, which as far as the history of human spaceflight goes is peanuts. It's also peanuts in relation to aviation, given the $330 million list price of an Airbus A380 (of which Virgin Atlantic have six on order).

It has to be said that *SpaceShipOne* was a 'suborbital' vehicle that took only short hops into space, but nonetheless the price was impressively low. It was built by Scaled Composites, whose operations are based in Mojave. They're now working on a small fleet of its successor, *SpaceShipTwo,* which will shuttle six passengers on two-hour flights beyond the lower atmosphere for a cool $200,000 a ticket. And Scaled Composites is not the only designer and builder of commercial spacecraft in Mojave. I also walk past hangars for Masten Space Systems and XCOR Aerospace.

XCOR's CEO Jeff Greason (whom I'll meet later) was part of the 2009 Augustine Commission that reviewed the workings of the American space program. Their report made the case for stronger partnerships between NASA and private spaceflight operators. Greason remarked that their findings departed from previous presidential commissions on the future of US space exploration not by recommending anything radically different but because 'some people actually paid attention this time.' Part of the reason, no doubt, is the Space Shuttle's 2011 retirement date. With no launch vehicle immediately to replace it, NASA must look to commercial providers to get its hardware and astronauts into orbit (if only to staff the International Space Station). Another reason was that *SpaceShipOne* provided a real-world demonstration of cheap commercial spaceflight.

A number of commercial spaceflight companies have popped up in the last decade, most funded by billionaires looking for a new challenge and/or some ego food. Some are beginning to do serious business. Los Angeles–based SpaceX (founded in 2002 by PayPal founder Elon Musk) has already secured launch contracts from NASA and commercial satellite developers. The company claims it has developed all the flight hardware for its *Falcon 9* orbital rocket, the *Dragon* spacecraft that is designed to sit on top of it, as well as three launch sites for less than the cost of one of NASA's launch towers – all while making a profit.

It hopes to be ferrying passengers to the International Space Station in the coming years.

Other start-ups are more interested in space-living. Hotel entrepreneur Robert Bigelow already has two 'inflatable' test habitats circling the planet and hopes to put a chain of space hotels and commercial residences into space, on the surface of the moon and, eventually, Mars. Before leaving for America I'd spent a mind-blowing afternoon with Xavier Claramunt of Barcelona-based *Galactic Suite*, who claims he'll have his first two paying guests in a space hotel by 2012. 'Our philosophy is the socialisation of space,' he'd told me over gin and tonics in the bar below his office. 'In fifty years we'll have sixty hotels around the world in Low Earth orbit. Space is nearer than people know. In the next fifteen or twenty years, space will be like going to the seaside. I am sure *you* will go to space.'

Xavier is also team leader of the Barcelona Moon team competing for the Google Lunar X prize, an international competition for privately funded companies 'to safely land a robot on the surface of the Moon, travel five hundred metres over the lunar surface, and send images and data back to the Earth' (a competition in which over twenty teams are taking part).

There's a new excitement about space again, and a lot of it has been generated by people working in this dusty collection of huts and hangars in the California desert. As Jeff Greason says, 'Mojave is to the emerging space industry what Silicon Valley was to the computer industry.'

My first stop is to see Stuart Witt, general manager of the spaceport. I'm hoping he can give me a measured view of the burgeoning commercial spaceflight industry. Like any frontier, commercial spaceflight is characterised by big personalities willing to take large risks, and the hyperbole can become intoxicating. Stuart, I'm told, is one of the more grounded figures, a man well respected in the industry who knows most of the players in it and provides services to many. He's also one of the founders of the Commercial

Spaceflight Federation (a body for discussion of standards, best practices and safety) and as such is not wedded to any company's particular story or ambitions. I feel I need someone who can help fortify me against any overly-optimistic resurgence of my childhood dreams of being an astronaut – or at least the idea of leaping on an economy flight to Earth orbit in the near future.

Stuart's office looks out onto one of three runways, and during our conversation light aircraft take off and land at regular intervals. On the walls are signed pictures of astronauts and aviators, interspersed with hunting paraphernalia. An ex-navy pilot, Stuart has a no-nonsense manner about him. He approaches conversation the way combat pilots approach missions – find the target and eliminate it. If he wants to make a point, he won't pussyfoot around. This isn't to say he's rude, in fact he has genial charm. Just don't waste his time. He tells me, 'The only people I care to be around are the ones that *truly live* the life they're given.'

It turns out Stuart's own ambitions were to become an astronaut, but despite flying fighter jets, he 'didn't have enough science,' for the training. Which is maybe why he's here. He still intends to get to space, although he's wary about the time frame. 'Space is still difficult,' he cautions. 'I would like to see hundreds of thousands of people get to space because I want to be one of them, but it's not going to happen overnight.'

First off, the industry has to deal with the issue of safety. The space shuttle has a fatality rate of just under two per cent. That sounds low but if commercial airlines operated under similar conditions over *fourteen million* people would have died in air crashes in the United States in 2009 alone. The aviation industry, in co-operation with its regulators, has achieved the staggering feat of taking what looks like one of the most dangerous forms of transport and making it the safest. Advocates of commercial spaceflight argue that the same will happen in their industry, but it won't be instant. The early years of commercial aviation saw significant levels of fatalities. Of the first nine de Havilland Comets (the world's first commercial jet airliner) built, five crashed. Two failed to get off the runway in time, one broke up in a storm over India, one ditched in the Mediterranean and another plummeted into the sea shortly after takeoff from Rome.

Mojave has already witnessed tragedy. Less than a hundred yards from where I'm sitting, in the spaceport's Legacy Park, is a memorial to three employees of Scaled Composites who lost their lives in July 2007 taking part in a propellant flow system test for *SpaceShipTwo*. 'At this facility I give you permission to kill yourself,' says Stuart. 'Here's my only limit on that: I don't want you taking out the whole neighbourhood when you do it.'

The question isn't *if* there will be an accident, it's when.

What that means is, if you want to take a trip into space in the next ten years, not only will you need somewhere around $200,000 spare, you'll have to sign an equally heavyweight 'informed consent' document. 'The people who will regulate this industry will be the underwriters,' says Stuart.

Risk is part of the deal here, and part of the attraction. Like everyone I meet at Mojave, Stuart has a deep-seated belief in the need for human beings to push boundaries and to explore. Commercial ambitions are subsumed into a higher aim than a simple search for profit. Pretty much everyone I speak to sees access to space as a battlefront for saving the human soul.

'We are creatures that require exploration, it's in our DNA,' Stuart says. 'Humans require risk-taking, they require bold leadership that is willing to investigate the things they don't know, because every society that has gone safe has fallen. I was raised out west where the night sky was this gorgeous canopy of stars and you could dream big. Some of us have a burning desire to be part of something greater than ourselves. That's the reason I'm here. I don't see that as a conservative or liberal viewpoint, it's who we are as a species.'

Stuart's face is now alive with a kind of fire. 'Innovation lives here,' he says urgently. 'We have unrestricted dream-space. That's what we believe in.'

Suddenly I'm eight years old again, dreaming of spaceships and touching the heavens, wanting some of that 'unrestricted dream-space' for myself. This may just be an industrial park in a sandy field on one side of a sleepy desert town, but something is happening here that will change the way we see ourselves. *Homo sapiens* are finally going to get into space in larger numbers. We will become a space-faring species.

Over lunch in the spaceport's cafe, Stuart introduces me to Dick Rutan, one of the world's most respected test pilots. He's perhaps most famous for co-piloting *Voyager*, a plane designed by his brother Burt to fly around the world non-stop without refuelling. The journey took nine days and covered 26,358 miles on one tank of gas. For aviation buffs, Dick is a superstar: aviation's version of Bono. Two people ask for his autograph as we are talking.

Dick has a history of putting himself in dangerous situations. He's been shot down over Vietnam, crashed in the Arctic and had to eject over the English countryside (while worrying he was going to plough his failing aircraft into the town of Brandon, Suffolk). 'I had about a second and a half,' he says. 'If I ejected in that second and a half, I could live; if I ejected before or after, I would die; so I didn't have a whole lot of time. I gave the airplane a little bit of left trim to avoid the town as I was reaching for the ejection panels.'

Rutan is like a coiled spring. He is now over seventy but is still flying high-performance aircraft and you get the impression he'll be doing it in another seventy by sheer force of will alone. He swears prodigiously, almost spitting out the expletives. 'Stephen Hawking is a prick!' he announces. 'You're telling me I can't go faster than the speed of light yet there's a gravitational force in a black hole that's strong enough to stop light and turn it around? Bullshit! Einstein's theory of relativity? Shame on you! Bullshit! Never look at a limitation as something you ever comply with. Never. Only look at it as an opportunity for greatness.'

Listening to Dick Rutan is compelling in the same way watching a car crash is riveting. You kind of want to look away but something akin to morbid curiosity gets the better of you. I don't mean to imply he is some kind of psychological wreck (his mind is as agile as the planes he's asked to test). More that he embodies a particular clash of man against world that's both intoxicating and scary. He's aware this doesn't always endear him to people. When I ask him what drives him, he's disarmingly honest.

'Request for recognition. Ego.' A pause. 'Arrogance.'

He smiles.

'You know what fighter pilots use for birth control?' he asks. 'Their personalities.'

My next stop is XCOR Aerospace for a meeting with CEO Jeff Greason. XCOR interest me because it's one of the few start-ups in this industry that isn't funded by a billionaire. You can tell that from their offices. Well, office (singular). Jeff and I meet in a shabby windowless hut furnished to give that authentic frontier vibe. Some of the seats are from old airplanes. There may be one of XCOR's bijou rocket engines on a shelf, but simpler technologies like the vacuum cleaner are clearly not welcome here. As we talk, it starts to rain, the downpour drumming loudly on the corrugated metal roof.

'Not everyone has picked up on the David and Goliath aspect of XCOR,' says Jeff. 'Richard Branson is pouring more money into Virgin Galactic in a month than we've spent in our history. You've got NASA doing billions and billions of dollars of its thing. Elon Musk pouring all of his money into SpaceX. Amazon.com's Jeff Bezos is funding Blue Origin . . .'

And yet, despite their underdog status, XCOR keeps being mentioned in the same breath as its far better-funded rivals. Talking to Jeff you don't get any sense of ego. He's trying to run a business – it just happens to be one that wants to make spaceplanes. That XCOR exists gives me more of a sense of a real spaceflight industry putting down roots. There's something very 'regular Joe' about Jeff. He has the chunky moustachioed friendliness of your local car mechanic. Having served on the Augustine Commission, he can happily express informed views on national space policy, but he can also instantly swap into a conversation about 'propellant mix' or 'fuel tank slosh.' I also get the sense that Jeff has arrived at the place he's supposed to be.

'As an eight-year-old growing up in rural Oregon I found this book called *Door into Summer* by Robert Heinlein in the school library,' he tells me. 'The main character takes a military application, figures out how to make a marketable product out of it and starts a small company that sets up in a hangar in the Mojave desert . . .' He's got a smile on his face that's broader than the state he grew up in.

Jeff shows me around XCOR's hangar where the company is building the Lynx spaceplane – a two-seater craft no bigger than a navy fighter that will take off and land on a normal run-way, carrying one passenger or an experimental payload. I sit in a fibreglass mock-up of the Lynx's cockpit and realise these people aren't joking. They really are forging a new industry. I look out of the cockpit and see engineers poking tools into rocket engines while Bon Jovi plays over the radio. Less than two-hundred yards away, Scaled Composites is building *SpaceShipTwo* not as a one-off, but as an actual production spacecraft.

'The idea is that we always fly full,' says Jeff. 'It'll just be you and the pilot. We also have complete reusability, it's a "gas and go" model. That's never been done. We're trying to make travel to and from space economically sane.'

Jeff reckons his break-even point is just two hundred one-hour flights a year at ninety-five thousand dollars a pop. Given that the Lynx is a 'gas and go' spacecraft, he estimates each one can fly four times in a working day.

In one corner of the hangar is XCOR's experimental EZ-Rocket aircraft, a precursor to the Lynx spaceplane that can climb into the sky at a dizzying ten thousand feet per minute.

'Dick Rutan was one of its test pilots,' Jeff tells me.

'Interesting guy,' I say. 'He's out to disprove Einstein.'

'Sounds like Dick.' He laughs. 'He's what you might call a strong personality...'

The following night Jeff and I meet for dinner, an Ultimate Omelette at the local Denny's. 'Ultimate' is something of a misnomer. I refuse to believe that I have enjoyed the pinnacle of omelette-kind. I particularly refuse Denny's in Mojave the claim that it can deliver me that zenith.

Jeff picks up on a theme I've heard a lot in my time here: environmentalism. It'd be easy to characterise the risk-taking, military-minded, hunting, hard-living, swearing, frontier-loving personalities that I've been talking to as unthinking searchers of novelty who appropriate resources to satisfy their personal ambitions with little care for the ecological impact. But that would be wrong. While nearly everyone I've met is dubious about man-made global warming (Dick Rutan called it 'a pathetic fraud, totally absurd and basically criminal') all of them are strong advocates of sustainability.

Stuart Witt had asked, 'How would you explain our unsustainable use of fossil fuels for energy creation to an alien? It's rather barbaric. So doing this thing we now dub "green" is literally doing the right thing. This isn't about politics, it's about doing the right thing as stewards of spaceship Earth.'

'I can't support the industrial-scale reckless waste of the planet,' says Jeff. 'I like this planet. It's where all my stuff is. I think we should manage the planet better because like it or not we are in charge of it.'

For the people here, it makes no sense to damage your own spaceship. Getting into space is a mental frontier that asks us to consider the Earth as a vessel that is carrying humanity. While most of us understand that idea in abstract, widespread access to space will bring the message home. 'With space as a single frontier, the entire planet is one civilisation,' says Jeff. 'It transforms the mindset. That's the idea that keeps me going. That's the big one.' It's the complete opposite of the ambitions that characterised the space race. The spirit of Mojave has perhaps less to do with being 'the biggest, baddest kid on the block.' More about 'humanity, you need to get out more.'

A lot of people might argue that we don't need to go into orbit to make us reflective about our role as stewards of the Earth, but as access to space becomes cheaper it can't help but subtly shift our perspective as a species. There isn't an astronaut on the planet that came back without some kind of transformation in their thinking.

As my bus pulls out of Mojave I'm convinced that commercial spaceflight is set to become a reality. Driving around the airfield with Stuart I'd seen real spaceships poking their noses through hangar doors. 'I don't think you'll see the first commercial operations out of this field until 2015 at the earliest and probably 2018,' he'd cautioned, which is a prediction some time behind those made by eager-beaver space fans and industry PR companies. Nonetheless, it's still an arresting thought. By the end of *this* decade the number of people who have taken a trip to space will, if you'll excuse the pun, start to rocket.

It'll be a long time before the average Joe can take a trip, though – not many of us have a hundred thousand dollars going spare for a couple of hours of sightseeing – but if commercial space flight goes the way of aviation, I may fulfil a childhood dream before I die. If I come back here to take that flight, the town will no doubt be transformed. By then that gleaming spaceport of the imagination could be a reality and my friend from Mike's Roadhouse Café and Bar will have to find a new place to live if he wants to avoid the temptations and trials of a thriving town. Mojave could become a byword for glamour.

For now, though, my method of travel is slightly more mundane, although my destination is not. I'm off to Washington, D.C., to meet one of the most important inventors in history.

CHAPTER 8

A constant and complete intercourse

We should not only use the brains we have, but all that we can borrow. WOODROW WILSON

I live on a hill, and at the top of that hill is a small park. It's part of the lifeblood of an area that, to be honest, is a long way from being a prestigious address. New Cross Gate, London SE14 is to Beverly Hills 90210 what The Troggs are to Bon Jovi (which is one of the reasons I like it). On summer evenings the neighbourhood tends to gather in the park to play Frisbee, drink, flirt, watch the sunset. From our position in the suburbs of the south-east corner of the city, we can see as far as Alexandra Palace ten miles to the north and Chelsea Harbour six miles to the east. The Houses of Parliament, the London Eye and Wembley Stadium are laid out for our pleasure. As dusk descends, the cityscape begins to twinkle as seven and half million souls reach for a light switch. It's beautiful. The same hilltop position that makes our sunsets so glorious also made it the ideal place, in 1795, to locate an 'optical telegraph station,' one link in an eighteenth century information network connecting capital to coast.

Optical telegraph stations were invented in France in the 1790s by brothers Claude and René Chappe, who were obsessed with finding a way of communicating messages without using horses or pigeons. They adopted a simple but effective visual system using a set of three connected wooden 'arms' (two short arms jointed to a longer one between them) that could be placed in ninety-eight different configurations by using a system of pulleys housed in a hut underneath. Providing there was a clear line of sight between huts, a message could quickly be communicated over long distances in short order (especially if the hut operators had telescopes). Messages that took days to travel by horseback could be transmitted in tens of minutes.

Napoleon Bonaparte was a big fan and, being convinced of success in a planned invasion of England, asked a third Chappe brother, Abraham, to design a telegraph that could signal across the Channel. He built the French-side station at Boulogne but ironically it was the home-grown British telegraph system (which used a system of six large shutters rather than movable wooden arms) that brought the message of Napoleon's defeat at Waterloo in 1815 to London. This was twenty years after my neighbourhood of Plow'd Garlic Hill became Telegraph Hill with the arrival of our own signalling hut.

The telegraph was heralded as a force for world peace. The 1797 edition of the *Encyclopaedia Britannica* suggested that 'the capitals of distant nations might be united by chains of posts, and the settling of those disputes which at present take up months or years might be accomplished in as many hours.' As the relatively limited optical telegraph network gave way to the worldwide phenomenon of electric telegraph in the second half of the nineteenth century, the 'force for peace' argument was trumpeted again. In his brilliant history of the telegraph, *The Victorian Internet*, published in 1998, Tom Standage quotes a toast proposed by British Ambassador Edward Thornton upon the completion of the first viable transatlantic telegraph cable: 'What can be more likely to effect [peace] than a constant and complete intercourse between all nations and individuals in the world?'

Standage raises these issues when musing about the claims made for today's Internet revolution, concluding that such hopes for our latest communications network are equally overstated. 'That the telegraph was so widely seen as a panacea is perhaps understandable. The fact that we are still making the same mistake today is less so.'

But while the Internet (and its predecessors) have clearly *not* delivered world peace, are they *part* of the reason the world is getting less violent?

Yes, I know. I was surprised too. The world is getting *less* violent? Have you *seen* the news? But the figures speak for themselves and they speak volumes. It's an astonishing and underreported fact that violence is declining and has been for centuries. This goes against popular sentiment that the past was somehow safer and simpler – a time without nuclear weapons or helicopter gunships, with no violent movies or 18-rated computer games. Steven Pinker sums up this misconception when he describes it as 'the idea that humans are peaceable by nature and corrupted by modern institutions.'

In the last thirty-five years, anthropologists like Carol Ember and Lawrence Keeley have been scouring the archaeological record and studying tribal cultures with results that seriously question the idea that the trappings of civilisation corrupt us toward violence. Their conclusion is startling. As Pinker writes 'the romantic theory gets it backward: Far from causing us to become more violent, something in modernity and its cultural institutions has made us nobler.'

Rates of savagery (the likelihood of being killed by another human being) can run to sixty per cent in tribal societies and tend not to be less than twenty per cent. 'If the wars of the twentieth century had killed the same proportion of the population . . . there would have been two billion deaths, not 100 million,' concludes Pinker. 'Global violence has fallen steadily since the middle of the twentieth century.'

I'm so (pleasantly) stunned by this thought that I decide I had better check the figures. I find myself searching the databases of the World Health Organisation for causes of death, reading reports like the University of British Columbia's *Human Security Brief* and trawling through spreadsheets with file names like 'Armed Conflict Dataset 1946–2008' offered by Uppsala University's Conflict Data Program. It's oddly heart-warming if you can get yourself above the fact that every statistic represents the culmination of a fellow human being's hopes and fears wiped out by a bullet, bayonet or landmine.

Just 2.8 per cent of the world's population died as the result of violence in 2002 according to the World Health Organisation. Over half of those deaths were the result of suicide (self-violence), a further one per cent homicide and just 0.3 per cent due to conflict. The WHO figures for the previous three years show a similar pattern.

Reading *Human Security Brief* you discover that the number of annual battle deaths in interstate wars fell from over sixty-five thousand in the 1950s to below two thousand per year in the first decade of the twenty-first century. The document is full of not-in-the-news statements like 'from the beginning of 2002 to the end of 2005, the number of armed conflicts being waged around the world shrank 15 per cent' and 'notwithstanding the recent increase in terrorist attacks, the number of civilian victims of intentional organised violence remains appreciably lower today than it was in the Cold War years.' There's more: the 1990s, it says, was the 'first time more wars (42) ended by negotiated settlement than by military victory (23). This started a trend that accelerated in the new millennium. Between 2000 and 2005, 17 conflicts ended in negotiated settlements; just four ended in victory.' The authors conclude, 'On average over the past six years, more conflicts have stopped than started each year. There is no reason to expect this trend to continue, but nor is there any reason to expect it to be reversed.'

That final sentence points to a problem with the figures: they're only a snapshot covering a short period of recent history. And statistics about levels of violence are extremely patchy.

There is no reliable and consistent source of data for war deaths (nations in the grip of conflict tend to make bad record keepers). The methods by which data is collected and reported (and the political motivations for doing so) are regularly challenged. Recent research by the Institute for Health Metrics and Evaluation in Washington and Harvard Medical School suggests that in some cases war deaths may be *three times more* than WHO estimates.

All that said, the long view is that the underlying trend in violent deaths is steeply downward. Even if we double, triple or quadruple the recorded rates of slaughter for the last century we're still killing far fewer people per capita than our ancestors. As Steven Pinker says, 'We must have been doing something right. And it would be nice to know what, exactly, it is.'

My hunch is that one of the things we're getting right is becoming increasingly connected, and being so we find it harder to kill each other. It's a popular view and one that seems to make instinctive sense. I'm conscious, though, that belief and truth should never be confused. I call some notable academics to see if there is any research that proves or debunks the theory. The Department of Peace Studies at the University of Bradford tells me my question is 'flimsy' and 'impossible to prove.' The widely respected history professor and author of *An Intimate History of Killing,* Joanna Bourke of Birkbeck University, replies to my email with 'I feel really stupid, but actually have no idea.'

The best evidence I can find are articles that give specific examples of communications technologies which help antagonistic parties share and process information, talk to each other or improve decision making in a non-violent direction. So, my hunch remains just that, backed up by a few scattered examples and a simple and seductive logic: that talk, trade and the sharing of culture across boundaries joins our fates together, giving us less incentive to kill each other.

This is what American philosopher Robert Wright refers to as the 'nonzero-sum game,' by which he means 'just because I win doesn't mean you have to lose.' My local park can be seen as a nonzero-sum game. It's not just a nice outdoor space: for families without a garden it's a place for their children to run free, and for the rest of us it's a space we can share without fighting for it. The restoration of the park in 2004 was a local effort spearheaded by Malcolm and Jayne, whose win was a win for all of us. This 'nonzero-sum' ethos is a strong part of my local community. John, Catherine, Patricio and Stephen are spearheading the construction of a much needed local café; Bridget runs the summer festival; and Jules, Sara, Susan, Martin, Phil, Max and Rosie give us an arts festival every Easter. Your body is nonzero-sum game too – each cell benefits from every other: if your liver wins, the rest of you tends to as well. (The converse is also true, as you'll know if you've ever had a hangover.)

Wright's argument is that 'through technological evolution new forms of technology arise that facilitate or encourage the playing of nonzero-sum games involving more people over larger territory.' This technology includes agriculture, urbanisation and, of course, communications and manufacturing technologies. Technology helps us create a worldwide park and café. To make the point he says, 'If you ask me why am I not in favour of bombing Japan I'm only half joking when I say, "Well, it's because they built my car."' (It's only sixty-five years ago that America dropped two atomic bombs on the Japanese – an act that would seem inconceivable today.) Slowly we're playing a bigger 'nonzero-sum game' with more and more of us involved.

Before you get the idea that Wright isn't acknowledging the troubles we face, he states that his theory is 'not intrinsically upbeat, it can accommodate the existence of inequality, exploitation and war . . . all it tells you is the fortunes will be correlated for better or worse. It doesn't necessarily predict a win-win outcome.' Taken over all of history he argues that 'people have played their games to more win-win outcomes rather than lose-lose outcomes. On balance I think history is a net positive.' He argues that there is a growing moral sense in the world, that the majority of us now believe that 'all people everywhere are human beings and deserve to be treated decently regardless of race or religion.' While the dehumanising effect of ethnic or ideologically based conflict is still all too obvious, it is no longer, says Wright, the prevailing view if you take the world as a whole. 'You have to read your ancient history to realise what a revolution that has been. This was not a prevalent view a few thousand years ago.'

It's a desire to talk about our increasing interconnectedness and how this will impact on our future that finds me approaching a large house in McLean, Virginia, just outside Washington, DC. We keep hearing about the Networked Age, but is it a good thing? Is it one of the reasons we're fighting less? Perhaps the man I have come to see can bolster or crush my belief.

This is by far the most affluent neighbourhood I have ever set foot in. Mansion-sized houses sit amid immaculately kept lawns. Wide driveways accommodate impossibly clean cars gleaming in the sunshine. I'm still a good ten yards from the door when it opens in anticipation to reveal a tall, healthy-looking man in his mid-sixties, immaculate in a three-piece suit, with a welcoming smile framed by a well-kept grey beard. 'Hello,' he says and extends a hand. This is Vinton, known as 'Vint,' whose success has had a fundamental impact on the world. Vint is one of the men who *invented* the Internet.

As we take seats in the library (it is that kind of neighbourhood) he hands me his card: 'Vint Cerf. Google. Chief Internet Evangelist.' 'I tried for archduke,' he jokes. The job title, he tells me, is a reflection of the fact that he's 'spent decades trying to get more Internet built and persuading people it's a good thing to do.' It's a mission he's still on. 'It's something I still need to do because there's only one-third of the world's population online,' he says. Vint believes in the virtues of interconnectedness.

I give Vint my card. 'This is sort of silly,' he says, taking it. 'It's the twenty-first century and we're handing out little pieces of cardboard.' As our conversation progresses I'll learn that Vint has an eye for the silly. A recent invitation to a gathering of Catholics saw him arrive in traditional Spanish academic regalia (a costume given to him at one of many ceremonies attended to receive his eighteen honorary degrees). When asked what religion he represented, Vint replied, 'Geek Orthodox.' He wore the same outfit on his first day at Google. 'I didn't want the young Googlers to think I was some boring old fart, and if they did, I at least wanted them to think I was an *unconventional* boring old fart,' he chuckles. At all other times, though, you'll find Vint dressed in a sharp three-piece suit. 'It's an ingrained habit of not wanting to look like everybody else,' he tells me. 'I wear them all the time. I wear them on flights so if I lose my luggage I'm still properly dressed for any event. I remember having to meet the Bulgarian president and prime minister and the airline lost my luggage for four days. I got along just fine.' Vint Cerf: old school, meticulous, playful – and smart as hell.

I'm hoping Vint can give me the big picture on our increasing interconnectedness. After all, he was in at the ground floor of the Internet and now works on the top one. He's a man with a career-length view on the technology, which for a technology as young as the Net is about the longest view you can have. As a graduate student, Cerf worked under Professor Leonard Kleinrock, who in 1969 oversaw the first computer-to-computer message to be sent using the 'packet switching' method that underlies the Internet. Actually, it was two-thirds of a message. Another of Kleinrock's students, Charley Kline, hoped to send a three-letter message 'LOG' to a receiving machine (this being the code for logging on to that computer). The 'L' and the 'O' worked but the 'G' crashed the system. 'So the first message on the Internet was LO,' said Professor Kleinrock. 'Or "Hello," crash!'

Cerf and others went on to help create ARPANET, a US government research project led by Lawrence Roberts. The ARPANET is often mistaken for being the first incarnation of the Internet we know today, but Vint tells me, 'The ARPANET became one of several networks that were interconnected to become part of the fledgling Internet.'

Vint is regularly dubbed 'the Father of the Internet' – an accolade he says is 'just wrong' and the result of 'an overactive public relations group' at his former employers, MCI. 'It causes a lot of trouble, because there are a lot of people that deserve credit. It's really painful for me personally to hear people say that and to have to try and defend against it,' he says. But if Vint isn't *the* father of the Internet he's one of them. It was Vint and his colleagues (notably Bob Kahn) who answered the question 'How can we reliably get data that lives on a computer on *this* network to another computer on *that* network when both might be using different hardware and software?' – a problem then called the 'Inter-Net problem.'

Their answer was a couple of 'software protocols' with the decidedly unsnappy title of 'Transmission Control Protocol/ Internet Protocol' or TCP/IP for short. (You may have come across this unfriendly couplet when you've been battling with your Internet provider.) On the Internet there is no fixed circuit set up between origin and destination along which data can pass. Instead, data is split up into numbered 'packets' which are sent into the wilds of the Net to find their own way to the destination. As each packet passes through any of the routing points on the network, it says, 'hey, I'm trying to get to B, do you know where B is?' and one of three answers will come back: 'Yes! *I* am B,' 'Yes, B is over there' or 'No, but I'm sending you to another machine who might know where B is.' When I check this summary with Vint, he tells me, 'Well, it is a *bit* more organised than that! But you are not far off!'

In a 2009 TED talk, Jonathan Zittrain, professor of Internet law at Harvard Law School, explained this process by asking his audience to imagine they were at a sporting event and somebody asks for a beer. 'It gets handed at the aisle and your neighbourly duty is to pass the beer along, at risk to your own trousers, to get it to the destination. No one pays you to do this. It's just part of your neighbourly duty. That's exactly how packets move around the Internet, sometimes in as many as twenty-five or thirty hops with the intervening entities passing the data around having no particular contractual or legal obligation to the original sender or the receiver.' Vint nuances this description. 'Actually most connections *do* involve contracts, even if no money changes hands, but the arrangements are entirely voluntary.'

At its heart the Internet is a system that works on trust and cooperation. It is itself an example of a nonzero-sum game. All systems cooperate, so all systems can benefit from the network. The software freely collaborates to split up, route and reassemble those packets of data so you can email your friends, store and retrieve files from the office, watch movies, video conference with your relatives in distant lands, and see Paris Hilton in all her night-time glory. 'If you look at the network itself, it is a vast collaboration, there is no central authority, and nobody is forced to build and operate pieces of it,' Vint says. 'I think that it should be self-evident that this notion of cooperation and mutual self interest is an extremely powerful part of the Internet story.'

At this juncture it's worth clearing up a common point of confusion – 'the Internet' and 'the Web' are not the same thing, despite often being referred to interchangeably. The Internet is deep plumbing. You won't really see it, any more than you see the sewer when you go to the bathroom. The Web, invented in 1989 by Tim Berners-Lee, sits on top of that plumbing to provide us with a service – a way to take packets of data and present them in a visual and interlinked way called 'web pages.' (The Web is not the only service sitting on top of the Internet, email is another.) The two terms, however, have become popularly interchangeable because it was the Web that suddenly made the Internet useful to a much larger audience.*

Beyond personal uses of the Internet/Web combo (being able to shop, get travel information, or find pictures of cats that look like Hitler) there are collective benefits too. 'One thing I will observe about the Internet which I find to be different from television, radio, the telegraph, telephone, newspapers and magazines is that it permits a kind of group interaction that has never been possible before,' says Vint. 'A larger number of people have the option of interacting with you – and I think that's important.'

The telephone (and its telegraph predecessors) allow one-to-one conversations, but those conversations can't cope well with

* When Berners-Lee wrote a memo suggesting a 'distributed hypertext system' to his boss, it was returned with the words 'vague but exciting' written on the front page. 'Eighteen months later my boss said I could do it on the side as a sort of a play project,' he recalls. That 'hypertext system' became the Web.

large numbers of people (as anyone who's been in a conference call will tell you). Video-conferencing suffers from the same constraints. Television, radio and newspapers are the exact opposite, reaching large numbers, but with little dialogue – they broadcast, we receive. The Internet/Web provides a platform to bridge that gap. Individuals across the world are able to form groups much more quickly and powerfully than at any time in history. Protest movements can achieve the critical mass needed to have their voice heard in a way that simply wasn't possible in a pre-Internet age. (Iraq war demonstrations used the power of the Internet to mobilise millions of demonstrators across the globe.) In between are groups of enthusiasts who collectively craft Wikipedia entries for the benefit of us all, or develop 'open source' software tools, or form online communities.

A nice example is The Legion of Extraordinary Dancers, who shot to fame in 2009 as an Internet phenomenon, going on to perform at the Oscars. They are basically kids dancing on street corners, who started filming themselves and putting their moves on YouTube. Kids from Detroit started innovating on moves created by dancers in Tokyo, and within hours new dance moves would evolve. The best dancers of the moment, with the craziest moves, created what is known as a 'spike' – a beacon that drew others to them, an audience they could never have reached before. Social networking sites buzzed with friends linking to videos of cool moves. *Look at this guy!* John Chu, who created The Legion of Extraordinary Dancers from finding the most popular of those YouTube clips says, 'Dance has never had a better friend than technology. Online videos and social networking have created a whole global laboratory online for dance.'

Marketing guru Seth Godin would call the Dancers a tribe:

> What tribes are, is a very simple concept that goes back fifty thousand years. It's about leading and connecting people and ideas. And it's something that people have wanted forever. Lots of people are used to having a spiritual tribe, or a church tribe, having a work tribe, having a community tribe. But now, thanks to the Internet, thanks to the explosion of mass media,

> thanks to a lot of other things that are bubbling through our society around the world, tribes are everywhere.

The argument is that the Internet helps us find our tribes, or take charge of a tribe that's been waiting to be led. ('The Beatles did not invent teenagers,' says Godin. 'They merely decided to lead them.') The Internet then is the great equaliser. You no longer need huge amounts of money or a broadcast network, you just need an idea and a leader whom enough people can gather around. It's not that the Internet makes us a global family, but it does allow us to create 'families' that cross boundaries of geography and wealth with greater ease, so many argue this can only be good for fostering a brotherhood of man. Beyond this the Net stimulates innovation, not just in dance moves but in science, technology and political thought. The Internet becomes the crucial engine that powers the information age – a 'Knowledge Combustion Engine' if you will.

But there's a dark side too. Increasing connectedness helps violent ideologues, paedophiles and bullies find each other (and their victims) too. Another problem is prejudice can be fostered rather than dissolved. In 2010, John Seely Brown, one of America's most respected thinkers on the interplay of technology and society said, 'The blogosphere and the echo chambers you find there are turning more groups violent. They're amplifying the extreme. You tune into the echo chamber you like best. There's no particular incentive to listen to ideas you don't like. We're going to have serious terrorism, self-generated in this country very soon, because of the Internet.' Robert Wright throws another element into the mix: 'trends in information technology [and] in technologies that can be used for purposes of munitions – like biotechnology and nanotechnology.' Those nanotech and biotech 9/11s I've worried about, after my visits to George Church and Eric Drexler, will be aided and encouraged into reality by Internet-enabled groups of extremists.

'In the United States we pay a very notable price for our freedom of speech,' says Cerf. 'It's hard for us sometimes to tolerate people saying things we don't agree with. But to preserve the society we live in you have to fight tooth and nail for that person's privilege to say what they just said. Not everyone in the world feels the same way about that.' The Internet exposes striking differences between nations. 'The European view of privacy and freedom of speech is different from the American version. But because the Internet is global and largely non-national in character it lands in the middle of all of these different, conflicting sociological perspectives.' Soon after my visit to Washington, Pakistan blocks access to Facebook and YouTube 'in view of growing sacrilegious content' demonstrating the tensions between politics, religion and free expression.

Reporters Without Borders, a non-profit organisation based in France that campaigns for press freedom, maintains a list of 'Internet Enemies' – as I write Burma, China, Cuba, Egypt, Iran, North Korea, Saudi Arabia, Syria, Tunisia, Turkmenistan, Uzbekistan and Vietnam. On its 'surveillance list' is Australia, which has proposed legislation to implement 'a draconian filtering system'. And as Vint and I sit talking, Google is battling with Beijing over the censorship of Web searches, a tussle that will result in the company shutting down its China-based search engine. It's another example of how interconnectedness doesn't always bring us together.

A different and interesting problem is that while violence is declining, we fear it more. Vint attributes this to the fact that today we 'experience more of the world than we would otherwise, and sooner, too – almost in real time.' This means 'we can misunderstand or misapprehend what it is that we have just learned or discovered or encountered.' The instant and sometimes unfiltered reporting of violence means 'you start to get the feeling that you are at risk, that the world is a dangerous place. Perspective starts to leak away because of this connectivity.'

To make the point he refers to riots in Kyrgyzstan, reported that day. In a less connected world, such violence in far-flung places wouldn't have even registered with us. Now, we hear about it wherever it is, and instantly. (The upside is that news of violence can 'get out' quickly. YouTube and wireless connections make brutality much easier to shine a light on.)

The idea that interconnectedness makes us less likely to fight is starting to take a few knocks, but none of them strike me as fatal blows. Certainly interconnectedness increases the instances of some kinds of violence, but for everything Robert Wright calls a 'death spiral of negativity' it's easy to find a League of Extraordinary Dancers, or a worldchanging.com (the influential hub that champions 'innovative solutions to the planet's problems') or a couchsurfing.org (a website used by millions that hooks up backpackers looking to sleep on a stranger's couch). If you look at the numbers, people use the Internet to connect with each other and learn. The world's most popular sites besides search engines are social networks, blogging tools and online encyclopaedias.

On top of that, the myth that the Internet is dominated by pornography also turns out to be just that. In 2006, Philip Stark at the Department of Statistics at the University of California, Berkeley, was commissioned by the US Department of Justice to look into the effectiveness of Internet filtering software. He concluded: 'About one per cent of the websites in the Google and MSN indexes are sexually explicit. About six per cent of queries retrieve a sexually explicit website.' That said, 'Nearly forty per cent of the most popular queries retrieve a sexually explicit website,' although Philip says that just because a search term may return a link to naughtiness that doesn't mean 'the person who ran the search was looking for it or visited it.' After all, lots of terms have double meanings, most obviously 'sex' which can refer to gender as well as horizontal athletics. The widely held belief that 'the Internet is awash with pornography' is perhaps better restated as 'the Internet is about one per cent pornography, but that one per cent is pretty popular.'

Back in Cerf's library our conversation turns to more upbeat territory. 'My optimistic statement of the day is not that information is power, but that information *sharing* is power,' he says. 'I think that's repeatedly demonstrated in the course of human history

– that the sharing of information makes us all more powerful – and that any society that suppresses information harms itself in large measure.' As Stephen Hawking said, 'With the technology at our disposal, the possibilities are unbounded. All we need to do is make sure we keep talking.'

The Internet is another forum in which to talk. Like the telegraph, the telephone or the newspaper, it's not perfect, but it adds to the number of ways we can converse to share information and ideas. 'The good side of it is that we encounter people we never would have encountered, we have an opportunity to rub ideas together we might never have had the chance to explore – and I think that's incredibly powerful,' says Vint.

Perhaps this is why staff at the Italian edition of *Wired* magazine nominated the Internet/Web for the 2010 Nobel Peace Prize and Vint Cerf, Bob Kahn and Tim Berners-Lee to be the recipients if it's accepted.

But connecting *people* is only a fraction of the Internet's story. As Cerf has written:

> In the next decade, around 70% of the human population will have fixed or mobile access to the Internet at increasingly high speeds. We can reliably expect that mobile devices will become a major component of the Internet, as will appliances and sensors of all kinds. Many of the things on the Internet, whether mobile or fixed, will know where they are, both geographically and logically. As you enter a hotel room, your mobile will be told its precise location including room number. When you turn your laptop on, it will learn this information as well – either from the mobile or from the room itself. It will be normal for devices, when activated, to discover what other devices are in the neighbourhood, so your mobile will discover that it has a high resolution display available in what was once called a television set. If you wish, your mobile will remember where you have been and will keep track of ... objects such as your briefcase, car keys and glasses. 'Where are my glasses?' you will ask. 'You were last within ... reach of them while in the living room,' your mobile or laptop will say.

'There's the "Internet of things" in addition to the Internet of people and ideas,' says Vint. As computing technology continues to get smaller, almost every object has the potential to become a node on the Internet. It's a world where if you've lost your keys you'll Google them, where your fire alarm will call you to let you know it's been activated, and your toothpaste orders more of itself as you reach the end of the tube.

'I think that "everything is connected" will not only be a technically correct statement in the mechanical world as, over time, more and more things literally are connected, but it's *always been true* in the conceptual sense,' says Vint.

It turns out Vint and I are both fans of James Burke, the British science historian and one of the faces of popular science on TV in the 1970s and '80s. Our favourite show was *Connections*, where Burke teased out links between society and technology. 'He's moderated some sessions at Google,' Vint tells me. 'He starts out with some fact and then he wanders around in 'connectivity space' taking you far afield and finally completes the loop back to the original point. I had a good deal of fun with him.'

Early in the first episode of *Connections*, first broadcast in 1978, Burke says 'The story of the events and the people who over centuries came together to bring us in from the cold and to wrap us in a warm blanket of technology is a matter of vital importance. Since more and more of that technology infiltrates every aspect of our lives it's become a life support system without which we can't survive.' Is the Internet the latest part of that ever growing support system? Certainly it's hard to imagine how many economies would cope if it suddenly stopped working. Physicist and computer scientist W. Daniel Hillis has written, 'We are now all connected, humans and machines. Welcome to the dawn of the Entanglement.'

I put this to Vint. 'This is not new,' he says. 'We have always been entangled with our technology, we've always been entangled with knowledge. It may be more obvious now, because of the way it manifests. But if you were a cave man you might have become quite dependent on tools that you built, because without them you might not be able to feed yourself, so you needed the knowledge to make those, or you needed the knowledge to find somebody who could make them. And then you also had to know that *that thing* over there was a sabre-tooth tiger and it was a really good idea to get away from it. The people who didn't understand that didn't survive.' In short, entanglement with knowledge and technology keeps more of us alive. The more connections the better.

But there's another worry. If the Internet is now, as Burke foreshadowed, a technology that 'infiltrates every aspect of our lives' and 'a life support system without which we can't survive,' what happens if someone switches it off?

In 1909 E. M. Forster wrote a short story called 'The Machine Stops.' In it, a worldwide society entirely dependent and cocooned inside an omnipresent machine meets its end when the machine breaks down. Millions die and the human race is left to a handful of forgotten humans who still live on the surface of the Earth. Will they restart the machine? 'Never,' says the main character shortly before he joins 'the nations of the dead.' 'Humanity has learned its lesson.'

Cerf is a big fan of the story. Could the Internet stop? Well, not easily. 'Parts of it become inaccessible, parts are under attack, information disappears because somebody shut down the website, or disappears because we don't know how to interpret the bits any more, pieces of it peel away and disappear,' he says. 'But I don't think you could stop the whole thing very easily.'

The reason for this is that the Internet isn't *a* machine, it's *billions* of machines. There is no 'off' switch. Its very architecture, its decentralised, nonzero-sum collaborative fabric means that the Internet is a bit like the planet's population. Or as Kevin Kelly, one of the founders of *Wired* magazine, calls it: 'the largest, most complex and most dependable machine we have ever built'. Short of an apocalypse you couldn't kill all of it, any more than you could wipe out everyone on Earth.

'Is there *anything* that could shut down the Internet?' I ask.

Vint considers. 'If every internet service provider in the world decided one day just to shut down the routers, that would pretty much screw the Internet,' he says. 'So the answer is, it's *technically* possible but it would require cooperative action that's extremely unlikely.' There is no conclusive count of the number of internet service providers on the planet but the figure is likely to run into tens of thousands. 'So,' asks Vint, 'now the question is, is there *hostile* action that could shut the 'Net down? Well, there *are* hostile actions going on every day all the time and they're capable of rendering *parts* of the 'Net inoperable but I don't think the machine would stop in and of itself.'

Technology's story is our story. Burke's 'warm blanket of technology' isn't separate from us, we're woven into the fabric of it and vice versa. And in the next chapter of the Internet's story, intertwined with 'the Internet of things' is something called 'augmented reality,' a phrase that strikes the same fear into my heart as those thin yellow burger slices that are 'cheese flavoured' and not *actual cheese*. What's wrong with *real* reality then?

A man walks into a shop and picks up a packet of paper towels. As he does so, an image appears on the packet telling him how much bleach was used in its manufacture. He picks up another and compares. The second gets a 'green light' that appears as a ghostly image on the back of the packet, signifying eco-friendliness. He chooses the latter as his purchase. Now he makes a phone call, holding out his hand where the numbered buttons of a keypad appear sketched in light on his palm. He dials the number by tapping his own skin. Later, on the way to the airport, he pulls out his boarding pass, and text appears telling him his flight is twenty minutes delayed.

This sounds like a scene from a (rather dull) sci-fi movie, but it is not. These are scenes from a demonstration of a new device developed at MIT's Fluid Interfaces Group – a combination of mobile phone, wearable camera and tiny projector that the lab's director, Pattie Maes, calls 'SixthSense,' a technology designed to provide seamless and easy access to 'information that may exist somewhere that may be relevant, to help us make the right decision about whatever it is we're coming across,' to help us 'make optimal decisions about what to do next and what actions to take.' Vint thinks it's great. 'It's a good example of what happens when your computer participates in the same real world that you do,' he says, smiling.

Technologies like these start from the assumption that every object is surrounded by a 'virtual' cloud of data you can't see. A piece of clothing isn't just the physical garment. In the shadow-world of data it is also how much it costs, whether it was manufactured ethically, the instructions for how best to wash it and so on. Crucially, your behaviour toward it may alter with access to any one of these pieces of data. Imagine walking into a trainer shop and being able to see from a product's bar code whether it was made in a sweatshop or not, or if the shop around the corner had the same shoes on sale.

In 2010, Microsoft employee Blaise Agüera y Arcas demonstrated the ability to link up online maps with photographs and video, allowing you to take not only a virtual walk around an area (including 'walking' *inside* buildings) but also to see what's going on at that moment, with real-time video links embedded into the scene. The same technology can be used to link historical photos and videos into the map, allowing you to step back in time. You can look into the sky and see star maps, or find out what blog entries refer to a particular place.

While these two examples are at the cutting edge of 'augmented reality' the layering data on top of our day-to-day experiences is already with us. Download the 'Better World Shopper' app onto your iPhone, for instance, and it will give you an instant rating of a manufacturer's record in regard to human rights, environmental policy, animal rights, social justice and community involvement. 'Google Goggles' makes use of your mobile phone's camera to recognise landmarks, book covers, even wine labels, and return Internet searches that relate to what you're pointing it at. This is data layered over reality, or (depending how you look at it) reality revealed by data. So, nothing like fake cheese after all.

'This ability to aggregate large amounts of information from many different sources in a coherent way has dramatically changed our ability to understand our world around us and to be aware of it and react to what's happening,' asserts Vint. 'Although I'm pretty sure that you could make a credible argument that these capabilities were present and have been present for a very long time in other media, they haven't been there in the same quantity, they haven't been there with the same immediacy and they have not been there to the same degree.'

'We are Entangled?' I ask, referencing Daniel Hillis's quote.

'Yeah,' says Vint. 'We are Entangled.'

A connected world is a good thing, then, but it hasn't escaped most people's notice that we're being told the planet is facing a crisis that overshadows all others. If the worst predictions of climate change come true, there won't be many of us left to connect, and despite all the civilizing effects of connectivity, violence will rocket as wars over land and resources explode across the planet.

James Lovelock, the scientist who originated the famous *Gaia* hypothesis of the world as an integrated and carefully balanced system, has said that if it warms up as much as he thinks it might in the next hundred years, 'We'll be lucky if there's a billion people left' (that means a cull of roughly eighty-five per cent of humanity). Others suggest the warnings of climate change are a grand hoax or a scientific orthodoxy that's become too powerful to argue against.

But this is clearly a big issue for the future and something I can't leave out of my investigations, so I'll soon leave America to visit the president of a nation that will entirely disappear if sea levels continue to rise and global warming makes good on the threat we're being warned about. But before I do, I'm heading to the Big Apple to meet the American oceanographer who coined the phrase 'global warming' and the German physicist who he thinks has a large part of the solution.

PART 3

EARTH

CHAPTER 9

World leaders still don't get it

A change in the weather is sufficient to recreate the world and ourselves. MARCEL PROUST

Climate change. I've tentatively raised the topic in numerous bars across America and the reactions have been fascinating. Some people – like my new astronaut friends – launch into a polemic about how it's all made up, while others do the exact opposite, angry that their fellow countryfolk are blind to risks that seem obvious. A small minority shrug their shoulders and say, 'I don't know what to think.'

But one thing almost everyone can agree on is that carbon dioxide levels in the atmosphere *are* rising. Ever since a man called Charles David Keeling started taking measurements of atmospheric carbon dioxide at the Mauna Loa observatory in Hawaii in 1958, the concentration of CO_2 in our skies has risen by about one quarter, from just over 315 parts per million in March 1958 to a notch over 392 in recent readings. Ice cores containing trapped bubbles of air from yesteryear allow us to estimate how much CO_2 was in the atmosphere much further back, and the record shows a sharp incline since the Industrial Revolution onward.

Another thing pretty much everyone agrees on is that global temperature has, on average, gone up by about 0.8 degrees centigrade since 1880. There is also no argument over the fact that the amount of CO_2 produced by man-made activity is rising. After all, our fossil fuels are largely made of carbon (coal being the most obvious example), so when we burn them to release energy, much of this carbon finds its way into the sky. Finally, historical records over much longer periods show carbon dioxide levels and temperature are generally in step with each other (though not always). So far it looks like a pretty conclusive case for the 'man-made global warming', then, doesn't it? Surely we can't expect to put a trillion tons of carbon dioxide into the atmosphere without something happening? If you're convinced by the climate change threat you're probably nodding your head, thinking 'good start' and waiting for me to bring in the big guns of massive scientific consensus, more than a century of laboratory experiments that show how carbon dioxide absorbs heat radiation and disappearing polar ice.

But I want to take a step back and ask why is it, for instance, that a national survey conducted in 2009 by the Pew Research Center (a deliberately non-partisan organisation funded by a charitable trust) found that thirty-three per cent of Americans don't think there is solid evidence for global warming and a further ten per cent can't decide? Or why is it that climate change scepticism is in political ascendancy in Australia, where public opinion shows a similar pattern? In the UK a number of recent polls suggest that anxiety about climate change has dropped (although the majority of citizens are still concerned) while the number of 'climate change agnostics' has risen to about a third of the population.

One of the reasons, surely, is that there are so many aspects of climate that we don't entirely understand. Take aerosols, for instance, by which I do not mean your underarm deodorant, but small particles put into the atmosphere by fires, volcanic eruptions, sea spray, dust storms, industrial processes and jet aircraft emissions. It's long been known that these affect the planet's temperature. The problem is we're not sure how much.

Yet they could be deeply significant. Aerosol particles provide something for water droplets to condense around and thereby form clouds (indeed, if there weren't any aerosol particles in the air there wouldn't *be* any clouds). The whiter the cloud, the more sunlight it helps reflect back into space, cooling the planet. (Everyone who's flown over banks of clouds will know how startlingly bright they can be.) In a 2006 paper published in *Science*, François-Marie Bréon at the Laboratoire des Sciences du Climat et de l'Environnement suggested human aerosol emissions 'may increase cloud cover by up to five per cent, resulting in a substantial net cooling of Earth's atmosphere.' Indeed, one technique proposed to offset global warming is a fleet of 'cloud seeding ships' that will scoop up seawater and force it through a system a bit like an inkjet printer to place tiny droplets of just the right size into the air, around which clouds can form. The argument is that more clouds could temporarily offset the heating effects of global warming, giving us longer to decarbonise our economies.*

We also know that some aerosols can reflect sunlight away from the planet by themselves. In 1991, the eruption of Mount Pinatubo in the Philippines sent roughly twenty million tons of volcanic ash twelve miles high into the atmosphere and average global temperatures went down by about half a degree centigrade the following year. The ice across Hudson Bay melted almost a month later than normal, and polar bears, who feed and give birth on the ice, had a greater number of healthy cubs that summer (offspring dubbed 'Pinatubo cubs'). Then again, other aerosols like soot *absorb* sunlight and have a warming effect. Some believe that this kind of 'black carbon' may be the world's leading cause of global warming after CO_2. Although on the positive side, as Mike Berners-Lee says in his book, *How Bad Are Bananas?: the carbon footprint of everything*: 'black carbon lasts only a few days in the atmosphere [so] if we reduce

* Nothing is simple in climate science. Some clouds actually heat the planet, notably high thin ones. It's the lower thick ones that reflect solar radiation. Inevitably, there is debate among climatologists about how the ratio of 'warming' to 'cooling' clouds will change with global warming.

the amount we create, the benefit will be instant. Hence some experts think that reducing black carbon pollution should be a number one priority in tackling global warming.' The problem is that we don't know how much of each sort of aerosol is in the atmosphere. In January 2010, a *Nature* article entitled 'The Real Holes in Climate Science' summed up the situation by saying estimates of the net aerosol effect vary by an order of magnitude.

Another gap in the climate change picture is understanding where the carbon we put into the atmosphere actually goes. We know how much we emit, and we know how much stays in the atmosphere (about half) but we're not sure exactly where the *other half* is going. It's certainly being absorbed by the land and seas but in what quantities and where is still not fully resolved. It was hoped that NASA's Orbiting Carbon Observatory would answer this question, except it crashed into the waters around Antarctica in February 2009. A replacement is on order, but with a price tag of $278 million it's not something that can be turned around in a weekend and for the moment scientists are making do with more limited observations from the Japanese GOSAT satellite.

So we know the amount of CO_2 we're putting into the atmosphere along with the science that proves it has a warming effect.* We're less sure of the impact of aerosols on temperature and exactly where the carbon is going. And we're even less sure of climate 'tipping points' that might suddenly trigger great leaps (or falls) in temperature. Which is a problem for people and politics. If someone's predicting a possible apocalypse we feel we ought to be one hundred per cent sure about it. We're told our cities may flood, whole nations may become uninhabitable, that the carrying capacity of the planet will plummet and wars will erupt as we fight for ever scarcer resources. Then we realise that climate science isn't exact. To the non-science literate this

* It's actually pretty easy to demonstrate the warming effect of CO_2 with a few test tubes, some stoppers, a couple of thermometers and ingredients you can find in your kitchen. You can make your own CO_2 by adding some acid (say, lemon juice or vinegar) to baking soda. Once you've made your CO_2, add it to some air in a test tube and quickly seal. Now seal another test tube without added CO_2 and stick both out in the sunshine for twenty minutes. Measure the temperature of each. Voilà! One is hotter than the other.

makes the whole enterprise sound rather unsure of itself – and given the gravity of the possible outcomes, many people expect a stronger-worded case. (Ironically for many scientists, the consensus on the climate change threat, expressed in a series of IPCC – Intergovernmental Panel on Climate Change – reports, represents the *strongest* wording any body of scientists has ever collectively come up with.)

Then there's the fact that the most important figure used in the climate change argument seems intuitively non-threatening – sure CO_2 levels have gone up, but by a hundred parts *per million*. And the rise since the Industrial Revolution? About a thousandth of a per cent of the entire atmosphere. As Australian MP Bob Katter said to his Parliament, 'Are you telling me seriously that the world is going to warm because there are 400 parts *per million* of CO_2 up there? If you know anything about science, you realise how utterly preposterous that proposition is, how absolutely ludicrous and ridiculous it is.'

Bob, unlike a scientist, or indeed a whisky drinker, finds it hard to understand that a small amount of something can have a significant effect. But he is not untypical among sceptics. And if you add to the uncertainties of the science, the fact that climate change is a slow-building problem that, if we accept and deal with it, requires fundamental changes to the way we organise our societies, it's not hard to see why for many people there is a very strong emotional reason to adopt a 'wait and see' attitude.

But just how wise is that?

About once a month I like to play poker with a few of my neighbours. It's an excuse to have a few beers, exchange stories and lose most of our money to Brian. We tend to play either Texas Hold 'Em or Omaha – games where an extra card is revealed on each round and players bet on how their hand is shaping up. I'm a 'wait and see' player – ever optimistic about the cards still in the deck. If I have a two and a five of hearts I'll hang on to the end, waiting for three more hearts to give me a flush. More often than not I should fold. I end up betting more money – and generally losing it. Brian is the opposite. If his hand looks weak, he folds early. He bets high when the odds are good, bails when they're bad – and occasionally, because we know this, he bluffs the hell out of us and wins with a pair of threes. (Bastard.)

My worry is that a 'wait and see' attitude in relation to the climate could be the worst mistake we ever make. Just because the climate models are incomplete doesn't mean that all those worried scientists are wrong. There are some pretty convincing 'known knowns' (to adopt Donald Rumsfeld's language). The temperature *is* rising. Arctic ice *is* receding. CO_2 concentrations keep going up. What if the IPCC's scientifically worded assessment that it is 'very likely' mankind is warming the planet becomes, 'in retrospect it was a certainty'? How much do we want to bet on this? Our major coastal cities? New York? Mumbai? Singapore? London? Tokyo? Or perhaps a few low-lying countries like the Maldives and Bangladesh? What about those billions of lives James Lovelock believes will be forfeited? Do we bet high or low? What are the stakes? Are we confident enough that those scientists are wrong? Are we willing to wait for the flush? I guess, at the end of the day, I'm not sure I'm happy to bet and then lose billions of my fellow humans to a world with a much lower carrying capacity and all the conflicts that will unleash.

From the air, you can see just how vulnerable New York is to sea level rise. As my flight approaches LaGuardia airport I find myself thinking that, to be convincing, sceptics actually need to show a *much stronger* hand than the climate change lobby. They need to *conclusively prove* global warming *isn't* man-made or isn't happening. It is not enough to *cast doubt* on man-made global warming. Sceptics need to conclusively nail that sucker. Because when it comes to the planet (as opposed to poker), I think I'd prefer to play safe. The worst that can happen if we're wrong about global warming is we lose some economic growth and accelerate the development of clean technologies to replace fossil fuels when they run out. The worst that can happen if the sceptics are wrong is that most of us die. In the words of *Dirty Harry*, if you're a sceptic 'you have to ask yourself one question: "Do I feel lucky?"' Well, do you?

But even if you are a sceptic, stay with me. The next stages of my journey will show me things that are good for the planet and our economies whether you accept man-made global warming or not. I land in New York as dusk is descending and take a taxi to the up-and-coming but knowingly shabby neighbourhood of Long Island City. I'm staying with arguably my most boffinesque friend, Colin, a neuroscientist whom I met via a flat-sharing website when I needed someone to share my rent in London. Colin is trying to get to the bottom of how memory functions in order to throw some light on diseases like Alzheimer's. This makes him a bit of a hero in my world.

Colin is actually in San Diego tonight being interviewed for a new job so I have his place to myself. Scattered about the apartment are research papers with titles like 'Hippocampal CA3 Output is Crucial for Ripple-associated Reactivation and Consolidation of Memory.' What's different about seeing this sort of thing today as opposed to when we lived together is that now I want to pick them up and understand them. (Previously, our scientific discourse stretched as far as designing the perfect breakfast sandwich.) One trip to Harvard, MIT and Eric Drexler's favourite coffee shop and I may be becoming a geek.

The next day is all sunshine and blue skies. I take a trip to Manhattan and sit overlooking the Hudson River in Rockefeller Park preparing for my next boffin-fest. I'm going to meet geoscientist Wallace Broecker and Klaus Lackner, a German physicist who may be to excess CO_2 what Dirty Harry was to criminals. I'm momentarily distracted by a boat emblazoned with the words 'America's *only* gay sailing tea dance.' Well, I guess it's a niche market.

Colin returns from San Diego in a dilemma. The job looks perfect, but he's not sure about leaving his beloved New York. 'I understand,' I say. 'There's no gay sailing tea dance in San Diego.'

'Right,' he says, and falls silent.

In climate science circles Wally Broecker is a combination of rock star and institution, the geoscientist equivalent of Willie Nelson. As I set off the next morning to meet him and Klaus, I'm feeling a mixture of apprehension and excitement. Broecker has been described as 'the grandfather of climate science' and 'one of the world's greatest living geoscientists.' He's the recipient of a shelf-full of awards, which if listed here would make your eyes glaze over, but include the US National Medal of Science and the Tyler Prize (awarded annually 'for environmental science, energy and medicine conferring great benefit on mankind'). 'I think the greatest pleasure is beating nature to one of her secrets,' he tells me when we meet. 'I'm an "inverse engineer" in a sense. We have an Earth system and I'm trying to work out how it's put together.' On his cabinet of gongs he remarks, 'Longevity helps.'

Among scientists Broecker is probably known best for his work on 'thermohaline circulation' (the ocean's 'conveyor belt') but more recently has become known as the first person to use the words 'global warming' in print. 'If my career has boiled down to *that* it's a big failure,' he says.*

My first stop is at Broecker's colleague Klaus Lackner's tenth floor office at Columbia University's Manhattan campus. In person Klaus is tall, with smooth features that if it weren't for his grey hair could fool you into thinking he was in his forties rather than his fifties. His facial expression seems in constant indecision about whether it is going to help him articulate a deep thought or a joke, which combine to give him an approachable air. Klaus is giving me a lift to meet Wally at his headquarters at the leafy campus of the Lamont-Doherty Earth Observatory, thirteen miles north of the city. The observatory is dedicated to studying the planet at a 'big picture' level, understanding Earth-wide systems and how they interact.

In the short time I have in Klaus's beat-up car it becomes abundantly clear he has a deeply analytical mind, inherited

* Wally remarks, 'That was in an article I wrote for *Science* in 1975. It was in the title so I guess any computer search can pick it up, I don't know if it's really true that I was the first one, but I certainly didn't use it with the intention of coining something – it just happened to be the likely thing to say.'

perhaps from his lawyer father who worked to build a fair judiciary in Germany after World War II. But as well as a scientist's need for clarity, there's an empathy for the ambiguities of the human condition. As we cross the George Washington Bridge, Klaus recalls in his perfect but slightly accented English the traumatic effects of conflict on his older relatives. Perhaps this gives some clue as to why Klaus isn't just an 'ivory tower' theoretician. Like his father, Klaus is determined to *do,* not just think. 'I have an engineering bent,' he says. 'So I'm not just looking at how and why things work, but how one can make things work. I'm very much interested in how to build things.' The world may one day be very glad that Klaus is like that.

I use the journey to ask how is it that a gas that exists in parts *per million* in the atmosphere can have such a dramatic effect on global temperatures.

'Well, in relative terms it's not small,' says Klaus. 'We've gone from two hundred and eighty parts per million to three hundred and eighty, so the increase is more than a third. If you drink three beers instead of two, that makes a difference. But the main reason it matters in the climate is because it's such a *potent* greenhouse gas.' In other words, a tiny amount can make a huge difference – a bit like putting a drop of Tabasco in your brother's contact lens solution.

'What makes CO_2 so potent?' I ask.

'It's a big player when it comes to infrared absorption in the atmosphere,' explains Klaus. Well, yes indeed. But what does *that* mean? It turns out it's all to do with wavelengths of energy. Which is a hard concept to visualise. We can see or feel energy's *effects,* but we can't see *it.* And given that energy itself is a pretty esoteric concept, talk of wavelengths of energy can be even trickier to grasp.

Most of us can, however, get our heads around wavelengths of *sound.* Nearly every hi-fi and radio has 'tone' or 'bass' and 'treble' dials to boost higher or lower frequencies. If you've got a really fussy hi-fi, you can fiddle with a 'graphic equaliser' to give precise volume boosts to the specific wavelengths you want more of (my brother, for instance, likes enough bass to challenge the integrity of his colon). We understand that low noises, like the bass riff from *Superstition*, have longer wavelengths than the masterful sax part, which we hear as a higher set of sounds (with shorter wavelengths). And we understand, too, that some materials absorb different wavelengths of sound better than others. The walls between me and my neighbours, by example, are pretty effective at taking out shorter sound wavelengths (high sounds), but not so good at filtering out the longer ones. As a result I'm intimate with the basslines of their entire record collection.

Klaus explains that CO_2 absorbs certain wavelengths of *heat energy* coming up from the planet's surface, just like my wall likes to absorb certain wavelengths of sound energy coming from next door. However, once CO_2 has absorbed energy, it likes to *re-emit* it, and a good proportion of that heat goes *back down* (to the Earth's surface) rather than *up* (escaping into space). The result? More CO_2 makes a warmer planet.

'CO_2 absorbs wavelengths of energy that other greenhouse gases don't,' says Klaus. To paraphrase an old beer advert: CO_2 reaches the energy wavelengths other greenhouse gases can't. In doing so it plugs some 'gaps' that usually let heat energy out into space. One of those other greenhouse gases is water vapour.

An atmosphere rich in water vapour is a large part of the reason we enjoy a habitably warm planet. The moon provides an example of what can happen when there isn't a lot of water vapour or other gases around to help form an atmosphere.

Water vapour is the largest absorber of heat energy coming up from the Earth's surface and is without doubt the most significant greenhouse gas. Its concentration in the atmosphere, however, is regulated by the planet's natural processes – and there's nothing we can do about it. If you pumped a whole load of extra water vapour into the atmosphere it would soon condense into rain and snow. Conversely, if you sucked some of it out, evaporation from the oceans would quickly replace it. The US National Center for Atmospheric Research estimates that every year 473 *trillion* tons of water evaporate from the ocean, and a further 73 trillion evaporate from the land. (This compares to only 30 billion tons of man-made CO_2 emissions.)

Going after water vapour as a way to combat global warming is not a possibility.

'The water vapour is uncontrollable,' explains Klaus as we drive north. 'It comes and it goes. But the CO_2? We put it in, and it stays.' (Or to be more precise, as David Archer, a University of Chicago oceanographer, puts it: 'The lifetime of fossil fuel CO_2 in the atmosphere is a few centuries, plus twenty-five per cent that lasts essentially forever. The next time you fill your tank, reflect upon this.')

There's a direct link between the temperature of the planet and the concentration of water vapour in the atmosphere. The hotter we get, the more water vapour is taken up by the air. This means that adding CO_2 to the atmosphere doesn't just warm the planet by trapping heat energy *in and of itself*, but encourages *more water vapour* into the air too. This means water vapour acts as a big *amplifier* of any warming effect CO_2 has. This is a key reason why CO_2 is so potent. When it gets into the atmosphere, it invites a whole load of warming water vapour to come with it (although it's *another* ongoing debate in climate science as to how much water vapour is taken up by the air as the temperature changes).

Klaus and I have arrived at the Gary C. Comer Geochemistry Building where Wally is waiting for us. The building is named after its benefactor, Gary Comer, founder of the Land's End mail order clothing company and a keen yachtsman. He had a particular interest in Arctic waters and first wrote to Wally in 2002 after navigating the famed Northwest Passage.

For the best part of four hundred years European nations (in particular Britain) launched scores of missions to try and find a navigable path through the Canadian Artic. The prize would be a sea route thousands of miles shorter to the Far East. In commercial terms the rumoured Northwest Passage was a prize worth dying for – and many did. There are stories of ships trapped in ice for up to five years, decimated crews limping back in damaged craft, the disappearance of entire missions and, it is now largely acknowledged, cannibalism. Ice thwarted nearly every attempt. Others were scuppered by madness, mutiny and politics before the ice could get them. In the end the passage was navigated in 1906 by Roald Amundsen, in the tiny, shallow-hulled *Gjøa*. Despite Amundsen's symbolic success, the Northwest Passage remained essentially unnavigable.

In 2001 Comer and his crew decided to see if they could take his 151-foot motor yacht Turmoil through the passage, expecting to fail (and having the safety of a sea plane on hand should they get into difficulty). Turmoil's crew sailed right through in just nineteen days with hardly any ice to bother them, which left Comer deeply concerned. Global warming, it seemed, was already making some very real changes to the planet. (Today, several large commercial ships have made the same journey. The ice barrier, for large parts of the year, is gone. For those who've studied climate change, or the history of the Northwest Passage, that is an incredible and incendiary fact.)

After his Northwest Passage journey, Comer decided to fund scientific research into the phenomenon and, after asking around the scientific community, was directed to Broecker. Under Wally's guidance he donated large parts of his fortune to a host of climate research projects – as well as putting up the money to erect the building I am now sitting in, just before his death from cancer in 2006.

If Wally were a Muppet, he'd be Waldorf, one of those old guys in the balcony – smart, funny and ready to point out, without remorse, the shortcomings of things placed in front of him. He has a charmingly curmudgeonly manner ('What are you doing here? You're writing a book? Oh yes, I have some vague recollection'), and at seventy-eight retains a good sense of mischief. He recalls playing a prank on fellow Lamont scientist George Kukla, jacking up his car and placing it carefully on cinder blocks just a mite bigger than the normal gap between the chassis and the ground. When George tried to drive off, his wheels spun impotently. The normally calm research scientist lost his cool, not least because he was entertaining 'a very important man from China, one of the first Chinese visitors that came here.' Wally laughs. 'We asked him whether the Chinese did pranks like that and he said: "Only small children."'

But joking aside, Wally is one of the world's top scientists, and when he talks about climate, people listen. He has kept himself apart from the Intergovernmental Panel on Climate Change committees, instead giving his own accounts, based on sixty years of science. He insists the warming we're seeing now is fundamentally different to historical shifts in the climate. 'It's bigger and faster,' he tells me. Which naturally prompts the question 'What can we do about it?'

The options generally presented are to do nothing (this from the sceptic camp), cut CO_2 emissions (from optimistic climate activists) or engineer countermeasures to produce some kind of counteracting cooling effect. This last option is called 'geo-engineering,' of which the wackiest idea is launching large mirrors into orbit. Geo-engineering is usually suggested by *pessimistic* climate activists.

But there is another option. Invest in Klaus Lackner.

Klaus and his colleagues have built a 'carbon scrubber' – a machine that strips CO_2 out of the air. Or to put it another way, on one side of Klaus's machine is air that contains current levels of CO_2 and on the other is air with roughly the same amount of CO_2 as was present before the Industrial Revolution. 'We spent the last five years in Tucson, Arizona, proving that this works,' says Lackner. The 'we' in question is Klaus and two brothers brought to the table by Wally: Allen and Burt Wright, who are unrelated to the famous flying Wright brothers, but might just get an equal place in engineering history. For if the scrubber technology is adopted, it can begin to *reclaim* the CO_2 we've been emitting, treating emissions in the same way we treat sewage. This could be a crucial component in a CO_2 processing infrastructure for the planet.

It isn't the whole or only solution (although with enough of Lackner's machines it arguably could be). 'If you've built a coal plant that captures CO_2 right out the smokestack, I can't compete with that,' says Klaus. But even if every power station suddenly became a zero-emitter of carbon tomorrow, there are plenty of other places pumping it out, especially in the transport sector, which accounts for nearly a quarter of the world's emissions. 'An airplane has a hard job running on electricity,' he says.

The Lackner/Broecker position is that creating waste isn't necessarily a bad thing. *Not dealing with it* is the bad thing. Nobody suggests you stop going to the toilet, but we did install sewage systems. In the UK it was actually 'The Great Stink' of 1858 (during which the smell of untreated sewage assailed central London, including Parliament) that finally convinced legislators to invest in sewers. You won't find many people arguing against sewers today but there were plenty of sceptics at the time, who saw public investment as 'ludicrous and ridiculous'. Lackner's carbon scrubbers make impossibly simple sense. What's more, they aren't just an idea on paper. Lackner's self-confessed 'engineering bent' means that, along with the Wright brothers, he's already built laboratory-scale working prototypes.

Mind you, carbon scrubbers aren't a new invention. They've been used for decades, for instance, in submarines to keep the air breathable. Until recently, however, the prevailing wisdom was that such scrubbing technology could not be adapted to remove the relatively small proportion of CO_2 in the atmosphere without using huge amounts of energy. Indeed, a 2005 special report from the IPCC dismissed Klaus's work with a single line: 'The possibility of CO_2 capture from ambient air (Lackner, 2003) is not discussed in this chapter because CO_2 concentration in ambient air is around 380 parts per million, a factor of hundred or more lower than in flue gas.' Or in other words, you'd be nuts to try and find the CO_2 needle in the atmosphere haystack. Anything that might work would take up too much energy (and thus add more CO_2 to the atmosphere than it removed).

Wally initially had reservations too. The first time he saw Lackner talk, he thought the German was nuts ('energetically nuts', actually). 'Then we had more time to talk, and I immediately tried to hire him.' Lackner set out to prove his methods could remove CO_2 using low energy levels, and Wally was right behind him. That's because Wally believes that a plan that only focuses on cutting emissions 'is going to kill us.'

'People say we've got to stop using fossil fuels,' he says. 'We look at it the other way. They're going to be burned. The world's going to need energy, the developing nations are going to use more energy, so we better damn well figure out what to do about it. We can't change over [to renewable energy] fast enough,' says Wally. 'The world leaders still don't really get it. That's why we need air capture.'

So how do Klaus's machines work?

If you've done a bit of chemistry, you might recall that sodium carbonate reacts with CO_2 to create sodium *bi*carbonate (baking soda). Well, in Klaus's machines there is a hanging gallery of strands of a 'sorbent' resin – impregnated sodium carbonate – which react with the CO_2 in the air flowing over them, the captured CO_2 helping to create baking soda.

Capturing CO_2, though, is only one half of the job. Somehow you've got to get the CO_2 *off* the sorbent if you want the apparatus to be reusable and therefore cost-effective. Restocking the whole shebang with a new supply of sorbent resin makes things prohibitively expensive and energy hungry. This is where Lackner's resin comes into its own, by doing something that even Klaus admits is counterintuitive. In the presence of water the resin changes its affinity for CO_2, shedding its recently collected bounty. The 'collection' reaction takes a reverse step. Sodium *bi*carbonate becomes sodium carbonate.

What this means is that if Klaus pumps water vapour into his machines, CO_2 from the sorbent will 'fall off' the resin, allowing the whole apparatus to be reused. Condensing that vapour allows the captured CO_2 to bubble out the top, in the same way CO_2 bubbles rise to the top of champagne.

There's a kind of sweet poetry to one greenhouse gas (water vapour) collecting another (CO_2). After all, one of the problems with CO_2 in the atmosphere is that it encourages more water vapour into the air, thereby amplifying the warming effect. Here, thanks to the chemistry of Lackner's sorbent, the opposite is happening. Water vapour is being used as part of a process to take CO_2 *out* of the air.

Of course, once you've captured the CO_2 you've got to store it somewhere (a process called 'sequestration') and make sure it doesn't find its way back into the atmosphere. One option is pumping the gas into rocks: specifically 'Ultramafic' rocks, which have high concentrations of magnesium that react with CO_2 to create magnesium carbonate. Klaus tells me Oman alone has enough of these rocks such that there isn't enough coal left to burn that would overwhelm them as a CO_2 store. Another option is pumping CO_2 into basalt rocks to make limestone – an alternative being investigated by another Columbia scientist, David Goldberg.

'Can the chemistry of the sorbent be improved further?' I ask.

Wally jumps in with a guffaw. 'They don't know how the chemistry works!' he exclaims with boyish joy, before pointing out a sealed tube (next to a can labelled '*Dr Fozze's Fart Beans*') that contains some of the first CO_2 captured by one of Lackner's early prototypes.

He's right. Klaus isn't sure why water vapour makes his resin give back some of its CO_2. 'I can tell you for sure *what* it does,' he tells me. 'That we can see. But at the moment I can only speculate *why* it does it. I've a good theory, and we will prove whether I'm right or wrong. One of the reasons I'm excited about where we are right now is we are setting up experiments to understand the chemistry. Once we've done that, we can *engineer* the chemistry. So I guarantee you these machines will get better.'

Klaus reckons that, once his scrubbers are in basic commercial production, for every twenty CO_2 molecules the machines put in the atmosphere (if they're powered by electricity generated from fossil fuels), they'll take out a hundred. And he's just at the beginning of his journey. With investment and experience, that ratio will get better (Klaus already has a long list of improvements he wants to research.) So it seems crazy that Klaus has struggled to raise the twenty million dollars he estimates he needs to turn his prototypes into a blueprint for a mass-manufactured unit.

'The fact that Klaus has trouble raising money is absurd,' says Wally, bristling. In fact, when I meet them, the two men are reeling from a recent decision by the federal government not to fund a research hub dedicated to carbon capture and storage. Klaus's requirements amount to just 0.000025 per cent of the $787 billion the US government pulled out of the hat for its American Recovery and Reinvestment Act of 2009 – money for stimulating the economy out of the economic crisis. One quarter of one ten-thousandth of a per cent.

It's ironic that when it came to saving the financial system, governments around the world couldn't move fast enough. Yet there is another platform all the banks run on. It's called the atmosphere, and the social and financial implications of global warming will do more to hamper Wall Street than anything they've done to themselves. When, I wonder, did a human-friendly atmosphere not become an infrastructural investment? A back of a napkin calculation suggests we could build enough scrubbers to reclaim all the carbon we pump into the atmosphere each year (and start to reclaim the backlog) for the equivalent of a three per cent tax on car prices for the next ten years.

Why does Klaus think it's been hard to find the necessary funds?

'I think our nature is that if the crisis is tomorrow we'll jump, we'll have the adrenaline to do whatever it takes to solve the problem,' he says. 'If I told you fifty years ago that what was happening in banking would lead to a meltdown ... would we have done anything? We are not good at thinking beyond a fifty-year time frame.'

Maybe we need another 'Big Stink'?

In the meantime Klaus can explore another avenue. CO_2 has lots of industrial uses. For instance, it's used to improve yields in commercial greenhouses. All plants require atmospheric CO_2 to live; it's the source of the carbon they convert into sugar and other carbohydrates through photosynthesis (and yes, this is another one of those feedback loops in the climate we don't fully understand – the extent to which adding more CO_2 will produce more plant growth on the planet). CO_2 also makes fizzy drinks fizzy, powers pressure tools, is a propellant for aerosol cans, and an ingredient in fire extinguishers. It can be used as a tool to create 'pure' atmospheres for welding, an aid for the construction of nanoparticles used in medical drugs, a refrigerant, a raw material to make polymers and plastics, and even as a theatrical effect (dry ice).

'The US consumes roughly eight million tons of commercial CO_2 a year,' says Klaus. 'So you could push this forward without government support. You start small, selling CO_2 into the market, improving your technology, and you can be ready before the coal plants have figured out the way to capture CO_2 at source.'

There's no way around it, Klaus is good news for the planet. Even better news is that he isn't the only one developing machines that take carbon out of the air. 'I convinced David Keith [renowned climate scientist] that this air capture stuff works so he now has a competing effort,' says Klaus. Peter Eisenberger, also at Columbia University, is attacking the problem as well.

Another promising idea is called 'Solar Thermal Electrochemical Photo Carbon Capture' that has been demonstrated by Dr Stuart Licht and colleagues at George Washington University. The team claim, if their technology was scaled up to cover seven hundred square kilometres, it could 'remove and convert all excess atmospheric CO_2 to carbon' in just ten years.

The more people working on technologies to take back the CO_2 we're putting into the atmosphere the better. Or as David Keith says: 'who will actually take it forward is now a horse race.'

'Let's imagine a world in which we suddenly have lots of Lackner scrubbers and you bring the levels of CO_2 in the atmosphere down to pre-industrial levels,' I say. 'Does the planet start cooling almost immediately? Does the warming stop?'

'We've warmed up the ocean and that will act as a damper on cooling,' says Wally. 'The ocean is holding back the warming of the planet too, because it sucks up a lot of the heat from the atmosphere. But as we cool the planet, the ocean's going to give that heat back, and slow down the cooling process.'

'The land would give its extra heat back in a couple of days,' explains Klaus, 'but the oceans will take decades, although you will see it going back down quite fast in the beginning.' Of course, we can't suddenly snap our fingers and fill the world with enough of Klaus's machines to start instantly offsetting all our carbon emissions. 'You can't do it overnight, but I do believe you could do it in a decade once you know what you are doing,' says Klaus. 'So, you'd have a thirty- to forty-year delay until you are back to normal.'

I ask Klaus how he goes about convincing people he's onto something.

'The problem I've found (and it's getting bigger all the time) is that I'm suspect to both sides of the debate. The people who make energy or are into coal feel I'm trying to stop them – because I'm saying you've got to take climate change seriously and business as usual is not acceptable. On the other side you have people who have some idea of what 'being green' means – and allowing people to use fossil fuels is not acceptable to them.'

'Do you think traditional environmentalists are part of the problem now?' I ask.

Wally snorts. 'Oh yeah!' Wally came into conflict with Greenpeace over their objections to testing deep ocean CO_2 sequestration. 'I get very upset. They strongly object to even *doing experiments*,' he says.

Is he optimistic we can solve the CO_2 problem?

'It'll be solved. The question is where will CO_2 levels get to before it's solved?'

Klaus agrees: 'I'm optimistic that ultimately it will be solved. But my view of human nature is that we will not solve it until we get seriously goosed.'

'Maybe in twenty years, when the impacts become obvious, we'll get serious,' says Wally.

'But let me give you an optimistic view,' says Klaus. 'Back in the 1990s I was asked, "How do you see this moving forward?"

and I said, "In this decade, the nineties, you will see scientists thinking about it and not much more. The next decade there will be a big political debate and not much more. The decade where steel starts to go into the ground is 2010 onwards. And people get really serious about it between 2020 and 2030." In a way, we are on that track.'

Wally announces that he has to go for a beer with George Kukla (the car prank obviously long forgotten), signalling the end of our conversation. Before he goes, he points to a picture on the office wall which shows him in a group of honorary degree recipients from Cambridge University. Standing near Wally is Microsoft founder Bill Gates.

'Why isn't he giving you some money?' I ask.

'I did send him some stuff but didn't get a reply,' says Wally, and heads off for his drink.

As Klaus and I make our way back to his car, we pass a huge furry pink and blue toy snake pinned to the wall outside Wally's office. Underneath it a piece of paper bears the words, 'I am the climate beast and I am angry!' This is arguably Wally's favourite metaphor. His assertion that 'The climate is an angry beast and we are poking it with sticks,' is one of the most quoted summaries of our problems with carbon dioxide (CO_2 being the 'stick' in question).

Klaus is giving me a lift to Dobbs Ferry train station for my trip back to Manhattan. As I get to his car I turn to him and say, 'You must be excited?'

'Oh yeah,' he says. 'Oh yeah.'

As my train makes its early evening journey along the east bank of the Hudson, something is scratching at the back of my mind. I know I've heard something on my trip about CO_2 that's significant to the conversation I've just had with Wally and Klaus but it's eluding me. I pass through Hastings, Greystone, Glenwood, and then, as we pull out of Yonkers, I remember: Joule Biotechnologies, a company George Church had mentioned to me in passing (and one of the many organisations to which he is a scientific adviser). The company uses genome engineering to make strains of bacteria that, under sunlight, consume CO_2 and excrete diesel and ethanol fuels. I whip out my iPhone and find their website, silently thanking Vint Cerf that I can.

'Joule efficiently captures sunlight to produce energy in liquid form,' it says, claiming its 'use of waste CO_2 as a sole feedstock creates the potential to deliver virtually unlimited quantities of fuel.' In April 2010, MIT's *Technology Review* named Joule one of the ten most innovative companies in the world, noting that it bypasses the major criticism levelled at biofuel production – the amount of crops and land needed to grow biofuel feedstocks such as corn, sugar beet or switchgrass.

Sugar beet and corn produce about five hundred and eighty gallons of ethanol per acre according to the Renewable Fuels Association, a lobbying group for the US ethanol industry. Estimates for algae biofuels vary wildly but even those on the conservative side suggest it can do about ten times better, with per acre yields of five thousand gallons of oil (which then needs to be refined). Florida-based Algenol Biofuels has developed a 'hybrid algae' that, via photosynthesis, converts CO_2 *directly* to ethanol. They claim a yield of six thousand gallons per acre and received one of fifteen $25 million grants issued by the US Department of Energy to help develop a pilot 'biorefinery' in Texas. Joule claims their system is *already* delivering six thousand gallons of ethanol per acre, and this figure will rise to twenty-five thousand per acre once they start full-scale commercial production.

Ethanol has struggled to get off the ground as a fuel largely due to costs of production (and linked concerns over land use), but with innovations like Joule's and Algenol's, that could be about to change. The US consumed 137,800,000,000 gallons of gasoline in 2008. If all of that had been sold via gas stations (there are 162,000 in the US) it would mean, on average, each outlet would have distributed just over 850,000 gallons that year. Assuming all gas stations are born equal (of course they're not) each one would need to find between thirty and forty acres of land to make enough fuel using a system like Joule's (this takes into account the extra ethanol needed to compensate for the fact that it delivers slightly fewer miles per gallon than gasoline). That's just one sixteenth of a square mile. Couple Lackner's CO_2 capture technology with a Joule or Algenol system, and you could build an ethanol or diesel factory nearly anywhere where there's sunshine. We could literally pull our fuels out of the sky. Of course, the burning of those fuels would put the CO_2 back into the atmosphere but, crucially, such a system wouldn't *add* any carbon dioxide, as our current use of fossil fuels does. It'd be what eco-geeks call 'carbon neutral.'

It might even appeal to those who don't buy the need for carbon neutral fuels. Today nearly ninety per cent of oil reserves are held by just thirteen countries, which makes the rest of the world dangerously dependent. The US, for instance, imported fifty-seven per cent of its oil in 2008. As reserves dwindle (and it's pretty much accepted that we will reach 'peak oil' at some point in the next generation), the opportunity for conflict is obvious. And beyond the possibility of making any country which successfully adopts these technologies energy-independent, they also eradicate the sort of man-made disaster that devastated the US coasts of the Gulf of Mexico as a result of the 2010 *Deepwater Horizon* oil rig explosion.

On top of all that, the energy infrastructure could stay largely the same. Gas stations would look like gas stations today; so would cars. There'd be no need to radically re-engineer our transport infrastructure as would be necessary in shifting to hydrogen-powered or electric vehicles.

Already there are eight million 'flexible fuel vehicles' on American roads that can run on a gasoline/ethanol blend (many are 'E85' vehicles that can run on blends of up to eighty-five per cent ethanol and fifteen per cent gasoline). General Motors has cautiously committed 'to making 50 per cent of production flex-fuel capable' by 2012. Ford, Chrysler and Toyota all offer E85 cars. In fact, Ford is returning to its roots – its famous Model T, which went into production in 1908, could run on ethanol, gasoline, or a blend of both.

So why not fuel made from CO_2 taken out of the sky? Already we have working technologies that can take CO_2 from the air and organisms that can turn CO_2 into liquid fuels. Klaus may not prevail; Joule might fail; Algenol might be a flash in a pan. I hope not. But even if they are, somebody will replace them. Of course it's not the whole solution. Capturing all the CO_2 from the transport sector and converting it back into carbon neutral fuel still leaves you with all those coal-fired power plants. But whoever prevails, as a model for partly weaning us off oil, it has to make sense. Even to certain Australian MPs.

I phone Klaus and ask him if the idea of using his CO_2 as a feedstock for fuels has occurred to him too.

'Yes, I've had some conversations along similar lines with some biofuel guys.'

'George Church?' I ask.

'No, but please introduce me,' he says. Which I do, by email, which feels pretty cool.

I've had an incredible day wandering, it seems, between apocalypse and salvation, and I feel the need for something ordinary and trivial to calm my mind. Colin comes up trumps. Along with a group of his friends, I'm taken on a tour of Manhattan's bars where we spend a large part of the evening comparing the relative merits of the Pet Shop Boys and Duran Duran. The latter, I suggest, were more fun and had better songs. Others disagree.

Sometimes, after a day of talking about things that really matter, you need an evening discussing things that don't.

CHAPTER 10

Here comes the sun (and it's alright)

Most of the shadows of this life are caused by standing in our own sunshine. HENRY WARD BEECHER

I started this journey wondering how long I'm going to live. Now I know: about ten minutes if Tracy Wemett has anything to do with.

There's a lot to recommend Tracy. She gives up her free time to mentor disadvantaged teenagers, organise food runs for the homeless and fundraise for an orphanage in Kenya. She's good at her job too – public relations. But her driving is without doubt the worst I've ever experienced. 'Nearly there,' she says before taking a corner at treacherous speed, while texting. 'There' could easily mean the afterlife. So I've never been happier to see a factory building in my life. The sign outside reads 'Konarka.' Tracy leaps out of the driving seat. 'I love coming here,' she tells me as I step tentatively onto the tarmac, relishing its ability to stay perfectly still.

From the outside, this nondescript building in an industrial park just outside New Bedford, Massachusetts, is nothing to write home about. Inside, however, something extraordinary is being made. 'This factory used to make photographic film for Polaroid,' says Larry Weldon, vice president of manufacturing, who's giving me a guided tour. 'That's why Konarka bought it. The process is similar. They bought the building, but they bought the machines too.' Larry accompanies Tracy and me along the production line. Huge spindles feed a continuous sheet of transparent plastic about two metres wide through a series of machines, each depositing thin layers of material onto the film.

'Really it's just a big printer,' says Larry.

Like a lot of engineers, Larry is a master of understatement. Because this 'big printer' prints rolls of lightweight flexible 'organic' solar panels just five millimetres thick that you can wrap around just about anything.

We've never had an energy crisis. A popular statistic states that the Earth receives more energy from the sun in a single hour than we use in an entire year. No, there is no *energy* crisis. What we have is an energy *conversion* crisis. Or more specifically, a *cost* of energy conversion crisis. That's why fossil fuels have done so well. Despite all their annoyances, they currently deliver more energy for your buck than other energy sources. That's because nature has taken millions of years to compress dead plants and animals into 'hydrocarbons' – materials made of only hydrogen and carbon.

Breaking the chemical bonds between atoms of hydrogen and carbon releases energy in the form of heat and light and is fairly simple (fossil fuels burn easily). It's because fossil fuels have a high density of molecules packed full of bonds ready for breaking that they release more energy when they are burned than, for instance, blancmange. Running your car on blancmange would require an awful lot of fuel. Another advantage of fossil fuels is they keep their energy nicely locked up until we need it, acting like a natural battery. Fossil fuels are the energy bank account Mother Earth has been paying into. Unfortunately we're making withdrawals from the account faster than it can be replaced. What nature took billions of years to save, we're spending in a matter of centuries. When it comes to fossil fuels, mankind is like an idiot with a credit card, and at some point the bank is going to cut off our funds.

No such problems with solar. The sun is waving a huge energy paycheck in our faces every second. It was the sun those ancient plants took their power from, using a process called 'photosynthesis.' It was those plants that long-dead animals (now, along with their food, turned into long chains of hydrocarbons) fed upon to get their energy. Why not bypass this billion-year waiting game? The problem is we've not been very good at banking solar, allowing most of it to slip away unused.

It's sometimes hard to credit why progress has been so painfully slow. Solar power has been a future icon since well before I was born. In his brilliant essay investigating its origins, 'The Beautiful Possibility,' Paul Collins quotes a *Popular Science Monthly* article: 'Future generations, after the coal-mines have been exhausted, will have recourse to the sun for heat and energy,' it announced. 'The sun will be the fuel of the future.' And the date of this article? 1876.

It was actually a translation of one that first appeared in *Revue des Deux Mondes*, a Parisian literary magazine that's still in business today. Back then France was marvelling at the work of Augustin Mouchot, who in 1869 used a parabolic mirror* to focus the sun rays on a copper boiler, creating a solar-powered steam engine; the same year he published the first book on solar power, *Le Chaleur Solaire et ses Applications Industrielles*. At the 1878 Paris Exhibition, Mouchot and his assistant Abel Pifre used a mirror over thirteen feet wide and a twenty-one-gallon boiler to power an ice maker – and received the exhibition's Gold Medal for their efforts. The next year, the exhibition was treated to a solar-powered printing press.

But despite patronage from Napoleon III, Mouchot's invention (unlike the telegraph) never caught on. High costs of

*Parabolic surfaces are shaped to collect energy from a distant source and focus it to a common point. Those big radio telescopes use them to collect and focus radio waves from the cosmos. Because the principles of reflection work both ways, you can also use parabolic surfaces to project energy from their focal point outwards, which is how car headlights work.

manufacture, coupled with the low price of fossil fuels, made it economically unviable – a combination that has dogged the solar energy industry for most of its history. But that's about to change – and, extraordinarily, Mouchot's method is at the heart of things. In a modified form and on a much grander scale, 'concentrated solar thermal power' is being adopted in the form of whole power plants – where banks of parabolic mirrored 'troughs' focus the sun's energy onto thin tubes of hot oil. The oil is used to heat water, which boils and creates steam to drive a turbine.

One particularly ambitious scheme is the proposed Desertec Industrial Initiative, which aims to put huge concentrated solar thermal plants in North Africa and ship electricity via massive power cables passed under the Mediterranean Sea to provide continental Europe with fifteen per cent of its electricity by 2050. Before you think this is idealistic posturing by eco-geeks, the partners in the project include Abengoa Solar (Europe's most successful solar power plant manufacturer), Deutsche Bank, Morgan Stanley and Siemens.

It turns out that our energy industries are staggeringly inefficient. To generate most of our electricity we take fuel out of the ground (usually coal) then drive it to a power plant where we burn it in large quantities to create steam, to turn turbines. The resultant electricity gets shipped along miles of cable to the end user. Coal plants lose over fifty per cent of the energy generated as waste heat and up to ten per cent of what's left can be lost in transmission down the wires (the figure was 6.5 per cent in the US in 2007).

For our liquid fuels we drill for oil in the few locations we can find it (less than 1,500 drilling locations account for 90 per cent of the world's known oil reserves) then ship this to refineries where it's processed before distribution to retailers. If you want any, you have to go and collect it (e.g., from a petrol station) or arrange a delivery. You then burn it, but miss out on most of the energy produced, losing up to eighty-five per cent.

While this all sounds ludicrously inefficient and inconvenient, it's still the cheapest way we've found to answer our energy needs. Solar power might sound like an obvious option but technologies for turning sunlight directly into energy have suffered from even greater inefficiencies.

At the same time, there's a clear case for improving the efficiency of the existing system rather than switching to new energy sources. While in California, I'd popped in to see Dan Reicher, Google's head of environmental projects. Dan argued that concentrating on energy efficiency is the best thing we can do for the planet in the short term. 'I apologise that this is not the exciting stuff and therefore may not make it into your book, but the low-hanging fruit is doing more with less; energy efficiency across the entire economy.'

Part of that solution will be the 'Internet of things' I talked about with Vint Cerf. 'We're starting to see smart appliances entering the market,' Dan said. 'The more they can talk to each other and your electricity supply about how much they're consuming, the more they'll be able to coordinate to use less electricity. Do I really care whether my dishwasher runs at six o'clock when it's a hundred degrees out and the electricity system is browning out because everyone's got their air-conditioning on, or whether it runs at three in the morning at half the cost with a lot less impact? I probably don't care as long as my dishes are washed next morning. We've developed the beginnings of some software where the grid can talk to various appliances in your house. Instead of turning on the next big power plant, the grid may say "we want you hundred thousand refrigerators not to go into defrost cycle for the next thirty minutes."'

The argument is that energy efficiency is the quickest and cheapest way to reduce the CO_2 load in the atmosphere. Short term, Dan is probably right, but making our use of fossil fuels more efficient doesn't change the fact that we'll still be using them (and using them up). We'll still be pouring CO_2 into the sky, even if we are doing it more slowly.

'This is one reason Google is developing new mirror technologies to help bring the price of solar thermal plants down,' says Reicher. 'We're looking at new and cheaper materials both for the reflective surface and the substrate that the mirrors are mounted on and hope to halve the cost of heliostats – the fields of mirrors that track the sun. Ideally we'd like to cut the cost by a factor of three or four.'

I'm convinced that concentrated solar thermal energy has a large part to play in our future, but I've come to New Bedford because what's happening here is part of the other half of the solar revolution – and it could see a radical shift in our relationship with energy.

Traditional silicon solar cells ('crystalline silicon cells') work because light energy hitting a silicon wafer messes with it, knocking negatively charged electrons off atoms. Each cell consists of two wafers, one placed on top of the other – one with added boron and one with added phosphorous. Mixing boron and phosphorus into silicon is called 'doping' and gives the wafers different electrical properties. The phosphorous-doped 'negative' wafer faces the sun, and when an electron gets dislodged in this part of the cell it will head off in the direction of the boron-doped 'positive' wafer underneath, setting off a chain reaction of 'charge passing' between atoms and creating an electrical current.

Silicon cells, however, are costly, complex to manufacture – and, ironically, have quite a high embedded carbon footprint (the 'doping' process, for instance, requires the wafers to be heated to slightly below the melting point of silicon). In recent years, 'thin-film' solar cell technologies have emerged that use much slimmer wafers of silicon or combinations of materials with catchy names like 'cadmium telluride' and 'copper indium gallium diselenide.' These cells can be made much thinner and lighter and are often cheaper than their crystalline silicon counterparts, which means that even though they generate less electricity per square foot, the cost of the energy produced can compare favourably. One (perhaps overstated) concern is that tellurium and indium are some of the rarest elements on the planet, raising worries that, ironically, these 'renewable' technologies will be hampered by the shortage of a natural resource. Another worry is that cadmium is one of the most toxic elements known to man.

Back in the Konarka factory, we've paused at a machine where the impossibly thin power-generating layer is being printed onto a roll of plastic passing through it. 'We don't use silicon, tellurium, indium, gallium or any of that,' says Larry.

'What do you use?' I ask.

'Plastic ink.'

Konarka's technology uses molecules of 'organic conductive polymers,' jointly discovered by the company's co-founder Alan Heeger, along with Alan MacDiarmid and Hideki Shirakawa (a discovery for which the three shared the 2000 Nobel Prize for Chemistry). The 'organic' label refers to the fact that the polymers in question are partly carbon. Because carbon is such a versatile atom (it will form all sorts of chemical bonds), it's the one nature prefers as the key building block for living things, which is why aliens in sci-fi movies of the seventies like to refer to humans as 'carbon-based life forms' (it's the first thing their scanners pick up). It's not just us, though. All plants and animals are carbon based. This is why chemistry involving carbon is often referred to as 'organic' chemistry.

Like the two types of doped silicon used in traditional solar cells, conductive polymer molecules can be either 'positive' or 'negative.' Like the phosphorous and boron-doped wafers in silicon cells, one set of polymer molecules gives up electrons and the other collects them. However, unlike the silicon wafers stacked neatly on top of each other, all these molecules are mixed up together in an inexpensive ink. The good news is that you can literally print solar cells. The bad news is that because the positive and negative molecules are all jumbled together, some of those electrons don't make it very far, getting captured back almost as soon as they've been released. So the material currently generates low levels of power measured in proportion to its surface area. Konarka's material, though, is much cheaper than silicon cells and can work in low light (and therefore for longer periods of time). In fact, before we'd started the tour, Larry had shown me a sample happily generating electricity under the fluorescent tubes dimly lighting the office where Tracy had introduced us.

Towards the end of the tour, Larry begins explaining one of the reasons Konarka's material (they call it 'Power Plastic') could be cheap to produce on a large scale. 'We can retrofit old photographic film plants,' he says, and although he won't confirm the figures, it's clear Konarka bought the building we're now standing in at favourable price. 'Polaroid wound up production in 2007. I was out of a job.' But not for long. Konarka bought the plant and rehired Larry and thirteen of his staff to refit the operation.

I'm looking at the fruits of their labours. Rolling past me are yards of film, all of it being turned into power-generating material. Its efficiency* might be low (currently between three and four per cent depending on the batch) but this plant looks more like a newspaper print-works than a solar cell factory. Konarka claims that in a single shift they can make a square mile of material.

I thank Larry for taking the time to show me around and begin to make my way out of the factory with some trepidation. I know the only way I'm getting back to my meeting with Konarka's CEO Rick Hess is with Tracy at the wheel.

Rick Hess is very CEO. He has the confident, relaxed air of a man who probably doesn't have to worry too much about his pension. His face is strong with big features: strong jaw, confident nose, eyes that meet your gaze full on. He doesn't so much

* Comparing the cost of energy produced by different solar technologies is difficult. Manufacturers like to quote 'efficiency ratings,' but this doesn't help much. If a solar cell is quoted as 'ten per cent efficient' it doesn't mean it will always convert ten per cent of the sunlight hitting it into electricity. A cell might be 'ten per cent efficient' in bright sunlight but generate nothing in lower light. Some cells don't make use of light coming in at shallow angles (early in the day or as the sun sets). On top of that, once silicon cells heat up past a particular temperature, their efficiency drops. Certain thin-film technologies by contrast show *increased* efficiency at higher temperatures. Taken together this means that a lower efficiency cell might deliver *more* electricity over time.

sit in the chair opposite me as own it, exuding pure-bred entrepreneur assuredness. Rick builds companies, sometimes sells them, and clearly revels in the corporate environment. He's also generous about his competitors. 'I'm a total believer that there are niches for everyone. Thin films certainly won't put silicon out of business, but they can be used in places where silicon can't.' He's also candid about the current limitations of his own firm's technology, notably its life span.

One of the major attractions of Konarka's technology is its flexibility, allowing it to be easily wrapped around anything from buildings to backpacks. Konarka even own the patent to solar thread, though they've not yet worked out how to hook up enough of the thread so the electricity can be put to use. One challenge they have faced is that, because the conductive polymers they use are 'organic', they quickly react with oxygen and moisture in the air. To keep flexibility but protect the power-generating material from these damaging reactions, Konarka needs to wrap it in a flexible airtight coating. 'The material we package lets moisture and oxygen in after three to five years,' admits Rick. 'We've tested ones that will give our technology a lifetime of over ten years but they're too expensive at the moment. We expect that to change.'

A major advantage is that Konarka's films can be made thin enough to see through, which means they can be used to coat windows and turn them into power generators. It's not a flexible application, but because the material can be put in the vacuum between double glazing it enjoys a lifetime of between fifteen and twenty years, allowing a building to generate power as long as there's light hitting the windows. It's one example of something your electricity supplier probably doesn't want you to think about – going 'off grid.'

'Phones went wireless and Internet went wireless and the only thing that's left that you have to find a wire for is power,' says Rick. 'You think about the developing world and what happened with communications. They skipped wires and went straight to wireless. I think for power they're going to do the same thing.'

A few months after my visit, the company began working with one of their customers to create solar-powered lanterns for developing countries, devices which store up charge in the day and can be used at night for study or work, as well as improving personal safety. 'Having light really does help a village and its people move to the next level of development,' says Rick. He's right. The village of Rema, a hundred and fifty miles north of Ethiopia's capital, Addis Ababa, has installed solar power. 'Our kids can do their homework at night now, because there is light. They are very happy,' says resident Elfenesh Tefera. Meanwhile the local bar can operate for longer hours and a solar fridge ensures the beer is always cold and the staff and customers no longer have to struggle with smoke from gas lamps. Since solar came to the village it's booming. It's the place to be with an influx of newcomers.

'Konarka co-founder Dr Sukant Tripathy's goal in starting the company was to ensure that developing countries were able to get power,' Rick explains. (The company was named in honour of the late Tripathy's favourite place, the Konark sun temple in Orissa, India.)

I ask Rick if he'd heard about a fuss a few months before which erupted when Xcel Energy, Colorado's biggest utility company, attempted to levy an extra charge on customers with domestic solar panels as a way to help pay for an expansion of the grid in the state. Rick laughs quietly. 'The utilities are doing the half-way model. They want solar power but they want to deliver it over the wires. There will be a place for that. Depending on where you live, it might make sense to buy your electricity from somewhere sunnier, where they can produce it cheaper.'

'Are you talking to utilities?' I ask Rick.

'No.'

'Because ... ?'

Rick doesn't answer directly, but tells a story. It's about a major telephone company sponsoring research into cell phones several decades earlier. 'In their grand wisdom, five years into the program they said, "This makes no sense! Nobody wants to carry their telephone around with them! We have wires all over the United States and people just go and pick up the phone and call. Why would they ever *carry* their phone with them? This technology is going nowhere."'

The implication is obvious. The days of an electricity system dominated by centrally-generated power are coming to an end, that solar could do for electricity what synthetic biofuels might achieve for gasoline: create a world where our energy sources are hyper-local. While I was at MIT talking to roboticist Cynthia Breazeal I'd taken the opportunity to meet with Bill Mitchell, head of the Smart Cities Group there. 'We're clearly going to see a movement away from old-fashioned centralised electrical grids to a much more decentralised system that looks a lot like the Internet,' he'd said. It's a nice idea, but there's still a huge distance to cover. Solar power remains expensive and there's a lot of inertia in our existing infrastructure. Rick agrees.

'I won't say when it'll happen, that's for the futurists to think about. We're getting on with making it happen. You just can't take the grid to everyone. It's not feasible.' He's right. Two billion people live without access to reliable electricity supplies. In Ethiopia, for example, eighty per cent of the population is rural, of which just one per cent has access to electricity. A truly national grid would be prohibitively expensive, which is why the country is experimenting with rural off-grid solar, as are Nepal, Sri Lanka, India, Vietnam, Ecuador, Tanzania, Indonesia, Kenya, Brazil, Ghana and numerous other countries. China is also forging ahead with ambitious grid-solar projects. The new world power is building a thin-film solar power plant in the Mongolian desert that's larger than Manhattan. When completed it should power three million homes.

A few months after I meet Tracy, Larry and Rick, something catches my eye. The town of Fowler, Colorado (the same state where Xcel got into trouble over its proposed solar power surcharge), announces plans to go off-grid with a combination of solar power, homegrown biofuel and gas derived from manure. 'The primary goal is to stabilise utility costs and then to reduce them,' says Wayne Snider, the town manager, 'but our ultimate goal is to become our own utility.' It's a hint of the hyper-local energy revolution Rick is talking about, in the heart of America.

'What about the storage problem?' I ask Rick. 'Solar only generates when the sun's shining.'

'Of course you need batteries to store excess energy generated in the day.'

One possibility is nanotechnology. As I found out in California, nanotechnology is infiltrating everything with the promise of driving down costs and improving performance – and the energy industry is no different. (Indeed, Konarka's conductive polymers are 'organic' nanotechnology.) Work at the University of Maryland on 'electrostatic nanocapacitors' aims to increase battery storage by a factor of ten, and they're already looking at integrating the results into solar cells. MIT researchers are even working on nanostructures that will hold charge indefinitely.*

Rick's solution is almost as revolutionary: a *printable* polymer-lithium battery. 'Our technology isn't limited to solar cells. You can make electrical components out of polymers too. So, for example, you could make a product with three layers: solar panel, polymer battery and organic LED light, all printed. A lot of hot countries need shade, so you make your shade out of this material. It keeps you out of the sun in the day and allows you to see at night.'

This is a key reason I came to Konarka, to get a glimpse at a wider revolution. Heeger's discovery of conducting polymers could change the way we manufacture electronic devices in the same way Gutenberg's printing press changed the way we made books. Which might explain why universities around the globe are setting up organic electronics laboratories at a rate of knots, many focusing on organic solar cells. A quick Internet search pulls up over seventy in countries as diverse as America, Germany, Japan, India, China, Morocco, Australia, Ethiopia, France, Canada, Czech Republic, France, Greece and Slovakia.

*Nanotechnology is also applying itself to improving the grid. Researchers at Rice University in Houston are working on ways to use super-conducting carbon nanotubes to do away with energy losses in transmission. The project, started in 2001 by Eric Drexler's former sparring partner, Richard Smalley, has created carbon nanotubes hundreds of metres long but only fifty thousandths of a millimetre thick. As yet, however, there is no way to mass-produce the right kind of nanotubes, but they're working on it.

In an example of just how fast the technology can be made and deployed, when the 2010 Haiti earthquake struck, Konarka employees used their weekend to build battery-charging units for Boston-based Partners In Health (a non-profit organisation dedicated to preferential health treatments to the poor), delivering them within twenty-four hours. PIH used the devices to provide power to places where there was no grid electricity so doctors and medical staff could work around the clock.

Each year the US Energy Information Administration publishes its International Energy Outlook, presenting several possible future scenarios for energy consumption including one where laws and policies relating to energy production stay the same as today. Like all energy predictions, it's based on assumptions and figures that you could argue with (and people do) but they're one of the few organisations who've had a crack at looking at the whole world (and drawing on my own experience of trawling through thousands of pages of energy statistics, I believe that alone merits praise).

It makes for grim reading if you're worried about man-made global warming. The 2010 report estimates that by 2035, world energy use will rise by forty-nine per cent from 2007 levels. While the use of renewable sources grow, so does the use of fossil fuels, with the result that CO_2 emissions go up forty-three per cent.

'Can solar scale up?' I ask Rick.

'Not all of the industry can, no. That's where a roll-to-roll process comes in. We can make solar material as fast as the spindles run and there's a lot of printing facilities in the world sitting empty. When the time comes we can scale up fast.'

While I know Rick is giving me the company line, there's already a precedent being set. People often quote Konarka in the same sentence as California-based Nanosolar, which also uses a roll-to-roll thin-film manufacturing process, coating their material with 'nanoparticle ink.' Nanosolar forgoes flexibility, mounting their material on a rigid substrate, but still claim they can produce a solar panel seventy-six inches long by forty-one inches wide with eleven per cent efficiency every ten seconds. Unlike Konarka, Nanosolar is concentrating on utility scale implementations. A few days before I meet Rick, the company announces orders totalling $4.1 billion from a number of solar power plant builders.

'Our scientists have goals this year to achieve two to three times the efficiency of our current commercial products,' says Rick. 'With biochemistry you can engineer molecules to do incredible things. For us it's a matter of designing polymers that both absorb more light and more frequencies of light,' says Rick. 'It's some pretty interesting chemistry but we're onto it.'

Again, it's the company line but I'm inclined to believe him. Not because of his winning smile and confident drawl but because I've seen what George Church is doing at Harvard in the field of synthetic biology. If Church can engineer living cells that make fuel, it's not hard to imagine that experts in synthetic chemistry like Alan Heeger can continue to push the boundaries of their science to make organic materials more efficient at generating electricity. 'We can also make materials that generate electricity from infrared heat radiation,' says Rick. 'So one of our patents involves a product that generates electricity from sunlight on its top surface but electricity from infrared on its underneath. The idea is you put them on the roof of buildings and use waste heat coming up, as well as the sunlight coming down to generate power.'

It's on the promise of advances like this that techno-optimists claim that widespread 'grid parity' (the point at which the price you pay for energy is the same whether it's been made by solar or fossil fuel) is but a hop, skip and a jump away. Others are less convinced, like self-confessed solar sceptic Dan Lewis, research director at the UK's oldest economic think tank, the Economic Research Council, though even Dan allows 'grid parity may become a reality in some sunny, major parts of the world by 2020.'

Rick announces he has to leave for a meeting but before he goes I want to ask him a final question. Rick is a guy who likes to make money. He's not only in solar for the good of the planet.

'Why are you doing this job?'

'I'm a disruptive guy. I like disruption. And solar is going to disrupt the energy market.' He pauses. 'But it does feel good to work on something that's a really good cause that everyone identifies with. My son is interested in what I'm doing. He never was before in his life! Dad's doing something that's cool!'

We say goodbye and Rick gives me one of those impossibly firm corporate handshakes that seems to say, 'Let's be friends, but know I will kill you if called upon.' Tracy turns to me.

'Need a lift?' she asks.

'You know what? It's sunny,' I say. 'I'll walk.'

Clean energy enthusiasts shouldn't bank on rising fossil fuel prices to stimulate demand for renewable sources. The oil and coal industries have a long history of finding new supplies, and while no one denies that fossil fuels will run out someday, this won't be what switches us to renewable sources in the time scale needed to halt our prodigious output of CO_2. Renewable sources have to get cheaper, and fast. And for a company like Konarka, this is a classic 'chicken and egg' situation. Its 'Power Plastic' material will start to get really cheap when production scales up to industrial levels, but until it *gets cheap* there won't be the demand for that level of production. (This is probably the real reason Rick isn't talking to utilities – right now his product is too expensive per watt to be of any interest to them.)

But the level of innovation in the solar industry is inspiring. There is no equivalent, for instance, of roll-to-roll manufacture in wind power. Thin-film manufacturers are springing up across the globe and each week seems to bring an announcement from one firm or another that their cells are pushing efficiencies up and costs down. Not all can succeed but some will, and I suspect it'll be the ones who exhibit the smartest use of nanotechnology.

Depending on which report you read and which year it was written in, solar energy is either the best-kept secret in the energy sector, showing exponential growth that will see it replace fossil fuels in twenty years, or a curious and insignificant sideshow. A more coherent, if counterintuitive, analysis is that it's both. Solarbuzz, a solar industry research and consultancy firm, estimates demand for solar energy 'has grown at about 30 per cent per annum over the past 15 years.'

But these figures come from a world where 'grid parity' has yet to be achieved. What happens when solar starts to get cheaper than the alternatives? It's not unreasonable to speculate that in a decade, solar will get much cheaper, possibly stimulating a radical shift to local off-grid energy generation, and achieve grid parity in many countries. Would we then see solar demand growing not by thirty per cent per annum, but fifty per cent? An annual growth rate like that would see solar quickly match all our needs, turning a sideshow today into the main feature tomorrow. It's instructive to look at the growth rate in Japan – one of the countries pretty much all analysts think will see grid parity in the near future. For 2009, Solarbuzz quotes a growth rate of 109 per cent.

Here comes the sun ... and I say, 'It's alright.' (Are you humming the fiddly guitar bit?).

CHAPTER 11
The black phantom

The best time to plant a tree was twenty years ago. The next best time is now. CHINESE PROVERB

I've come to the other side of the world – and it's very different here. Presenting myself to airport immigration, I'm expecting the familiar attitude of suspicion and disdain. I generally come away from national borders with the strong impression that most immigration officers are hired precisely because they have an all-consuming dislike of other people.

'What's the purpose of your visit?' asks the fresh-faced man behind the passport counter.

'I'm researching a book.'

'Oh really? What's it about?' comes the cheery response.

Is this a trick, the jolly demeanour? 'It's about the future,' I say carefully.

'Oh great!' he responds. 'I've got some ideas about *that*. I think solar is going to be really important.'

I'm taken aback and without thinking reply, 'It is!' before showing him the bag I'm carrying. Given to me by the good people of Konarka, it incorporates one of their organic solar panels and has been happily charging my iPhone since I left New Bedford. Before I know it, we're in a conversation about the role of technology in solving the world's problems. (He imagines a future of self-driving magnetic cars.) I look behind me expecting to see a queue of knackered and exasperated fellow passengers waiting to go home. Instead, they're all streaming through other channels exchanging friendly greetings with immigration staff. Have I entered an alternate reality? A border that deals with arrivals quickly and politely? One that actually *welcomes* you to the country? I'm already enjoying New Zealand and I haven't left the airport.

'Who are you seeing?' asks the guard.

'A lady called Vicki Buck.'

'Vicki's great,' he says and I'm taken aback a second time.

There again, it's not such a huge surprise. Vicki was the mayor of this city (Christchurch) for nine years, popular for her initiatives in tackling unemployment, invigorating the town's cultural life and addressing housing and social care issues. Today she's leading a charge for the renewable energy industry in New Zealand that could have international repercussions – initiatives that are undoubtedly good for the planet whether you worry about climate change or not.

The guard takes my blog address and promises to buy the book. 'Don't forget to mention those self-driving cars!' he says.

Vicki meets me at my hotel the next morning to drive us to breakfast. Although to say one 'meets' Vicki is something of an understatement. Vicki meets me in the same way a tornado 'meets' the air. You are instantly swept up, and it's invigorating.

The first thing you notice is she cannot stop laughing. She laughs at everything. It's not the laughter that comes from a good joke, more an all-consuming *joie de vivre*. Anything can raise a smile, animating her broad features into a mixture of delight and mischief. In her early fifties, Vicki has the energy of someone in their early twenties, but coupled with decades of practical experience that has emboldened her rather than beaten her down. Hard-won knowledge is married with an unbridled belief in the possible. In the week I spend here, I see Vicki wield authority without being authoritarian, command respect without demanding it and motivate action without telling anyone what to do.

In 1998, Vicki quit politics to concentrate on education, helping to set up two schools in the city in which she is still deeply involved. With these established, she began committing her almost inexhaustible energies to tackling the climate change issue. While she knows the state has an important role to play in dealing with global warming, her overall view is, 'if we wait for our governments to sort it out we're probably all fucked.' Oh yes, she has a reputation for straight-talking, too. This might explain why she preferred to stand for office as an independent rather than as a candidate for a political party. She's not wedded to any ideology other than the potential wrapped up in every individual. 'I like the "just do it" approach,' she tells me.

'Have you heard this song?' she asks dialling up a CD on the car hi-fi. The opening bars of a jaunty pop number issue forth. Vicki is tapping the steering wheel in time, the sun roof is down and for three minutes we listen to Kiwi songsmith Tim Finn's 'Couldn't Be Done' inviting us to prove all doubters wrong. It could be Vicki's theme tune. I get the impression it's on strong rotation in this vehicle. 'Isn't it great?!' she exclaims.

'I think you can achieve a lot more if you're enjoying yourself,' she tells me as we order breakfast, possibly explaining one reason why Vicki accomplishes so much. The list is frankly frightening. She runs through a small selection of her current interests: she's a director of New Zealand Wind Farms, of biofuel company Aquaflow Bionomic Corporation, the climate action website Celsias, and charcoal manufacturer Carbonscape.

It's this last company that has snagged my interest and the main reason for my visit to New Zealand. Charcoal has undergone a radical makeover in recent years. Creating it (and then burying it) is seen by many as one of the most promising ways to mitigate global warming. James Lovelock, who thinks we're heading for climate Armageddon, has even suggested, 'There is one way we could save ourselves and that is through the massive burial of charcoal.' In this context as a planet saver, humble charcoal has been given a flashy new brand name: biochar.

The biggest contributors to CO_2 in the atmosphere are plants and animals (both on land and in the sea). In fact, they dwarf human emissions. 'The biosphere pumps out 550 gigatonnes of carbon yearly; we put in only 30 gigatonnes,' says Lovelock. Sceptics often point to this difference and say 'See! Global warming has nothing to do with us!' Sadly, though, it does.

While it's true that the lands and oceans burp out massive tonnages of CO_2 they also take it back. In fact, it's a finely balanced system – what gets emitted gets absorbed again. This is called the 'carbon cycle.' Through the miracle of photosynthesis, plants play their part by taking CO_2 from the atmosphere and turning it into 'biomass.' Nearly sixty per cent of a plant's mass is carbon taken from the atmosphere. Most of that carbon is returned to the air when a plant dies – the vegetation rots and the carbon is released skyward. It's a natural cycle. Unlike our own use of fossil fuels, which has taken carbon locked deep underground (after slowly being buried and compressed over millions of years) and released it. The land and seas are taking up about half of this surplus. But half of it remains in the atmosphere.

At the heart of the great charcoal/biochar resurgence is a process called 'pyrolysis' – a way of burning biomass in a 'low oxygen' environment that creates a stable non-biodegradable form of carbon: charcoal. Unlike a normal open fire which turns what you burn into ash and CO_2, pyrolysis 'part combusts' it in an oxygen-starved enclosure, turning roughly half of it into charcoal (carbon) and the other half into wood gas and oil that can be used as fuel. ('Wood gas' sounds like something out of a Tolkien novel but it has a long history as an energy source. Vehicles powered by it were common during World War II when fossil fuels were rationed. Volkswagen and Mercedes had production models.)

Charcoal is locked-up carbon that, like coal, won't be released unless it's burned. It doesn't rot like the biomass it used to be before it was pyrolised. There's a neat symmetry to this. It's the release of carbon trapped in ancient buried biomass that's causing increased CO_2 levels in the atmosphere. Contemporary biomass is being used to retrap that carbon and stick it back in the ground.

Cornell University's soil scientist Johannes Lehmann (who shares the same campus as robot maker Hod Lipson) believes the potential benefits could be huge. Of the sixty gigatons of carbon taken out of the sky by plants, he estimates ten per cent ends up as waste in agriculture or forestry. Corn stalks, rice stalks, branch and leaf litter (as well as animal poo) are all potential food for pyrolysis, 'enough to halt the increase and actually decrease the level of atmospheric carbon by 0.7 gigatons a year,' he writes. 'Clearly, the potential contribution of biochar technology is large, perhaps large enough to mitigate climate change alone.' James Lovelock says this approach could 'start shifting really hefty quantities of carbon out of the system and pull the CO_2 down quite fast.' It's worth pondering that point. Lehmann is saying that enough waste biomass alone could, if turned into charcoal, not only offset all our carbon emissions but also start to bring CO_2 in the atmosphere *down,* reversing the effects of global warming.

It's important to note that Lehmann is one of the more optimistic voices when it comes to biochar. Others quote lower estimates for the amount of waste biomass available. But there's no denying billions of tons of it are produced each year. Beyond this, however, there is the issue of convincing farmers and foresters to become pyrolysis converts. Which is where companies like Carbonscape come in. Before I find out more about them, however, Vicki wants to introduce me to her latest obsession.

I'm looking at a system of municipal wastewater ponds on the outskirts of Blenheim, a small town in the wine-growing regions on the north coast of New Zealand's south island.

'I bring you to all the best places!' laughs Vicki.

Two nondescript blue shipping containers sit beside the ponds proudly bearing the Aquaflow logo. A man wearing Wellington boots, jeans and a T-shirt with the image of a goat's skull on it emerges from one of them. I instantly think 'engineer' and I'm right. This is Mark Vinsen, who explains what's going on.

'Water from the pond comes in,' he says, pointing to a pipe. 'Then we harvest the algae out of it.'

'Algae are very cool,' enthuses Vicki.

'Those algae have been feeding on contaminants in the water. By removing them, we're most of the way to turning wastewater into fresh clean water,' Mark continues. 'We put the water through a few more steps but the algae do the initial hard work.'

'Clean water is the major part of the business,' says Vicki.

Mark takes me to the end of the second blue container, where a green sludge that looks like a cross between pesto and snot is dropping off a roller. 'This is algae paste. We process it to create something we call "biocrude" that can be made into fuels.' (Aquaflow has manufactured synthetic components of aircraft fuel, for instance.)

'That's the other part of the business,' says Vicki.

I'm glad I've been distracted from charcoal for a minute. After learning about the potential of organism-made biofuels from George Church, it's nice to see the science made real. This isn't just an idea in a laboratory, here's a blue box in a field making the stuff.

'What kind of algae do you use?' I ask. I'm expecting to hear about a specifically engineered algal strain that Aquaflow owns the patent for.

'Whatever's in the pond,' say Mark. 'That's the beauty of it. There are thousands of varieties of algae and they vary depending on the time of year.'

'You work with what nature already puts in the pond? You're algae sluts?'

Vicki laughs. 'Yeah, when it comes to algae, we'll go with anyone! But what that means is that we can turn any municipal waste pond into a manufacturer of cheap clean water and biofuel.'

It's a pretty cool idea and one that will no doubt find its place. In fact, a few months after my visit to Blenheim, the company announces a partnership with the United States Gas

Technology Institute to 'demonstrate the conversion of algae biomass directly to gasoline and diesel fuel.'

We bid Mark farewell and drive off with Tim Finn keeping us company. 'How did you get into this?' I ask Vicki.

She laughs. 'I know! Algae? I mean I knew *nothing* about algae. But now? Now I love them. It's amazing what they can do. Seriously, I'm in *love* with algae. I'm a total algae bore!'

Twenty minutes later I'm in a shed on an industrial estate staring at the machine that's brought me halfway around the planet. It looks like it was designed on a drunken night out after a special effects conference. I imagine a tussle between the *Harry Potter* team and the *Star Wars* crew.

'It should have a big cauldron, right in the middle!'

'Well, if you're going to have that, we want a bank of computers on the side.'

'Okay, but it has to be *black*!'

It even has a fantasy sci-fi name.

'Behold "The Black Phantom,"' says Vicki.

This is the jewel in Carbonscape's crown: a black metal edifice about eight feet high, dominated by a huge cauldron-like pot, large enough to contain a man. In fact, as I approach, a man in dark blue overalls holding a screwdriver pops his head up over the rim. 'Hi,' he says. 'I'm Greg.'

We're joined by Carbonscape co-director Tim Langley, a man who seems to mix the countenance of lumberjack and monk. There's something about him that exudes acceptance of life. It doesn't take much for his face to smile, and when one comes it's not like any effort is needed, more that a light has been shone onto a grin that was there all along. I ask him to explain the Phantom.

'Well, essentially it's a big microwave oven,' he says.

'I see.'

'Inspired by a potato.'

'Of course.'

The machine is the idea of paleoclimatologist Chris Turney, professor of physical geography at Exeter University, UK As a teenager he baked a potato for forty minutes in the kitchen microwave creating a mass of potato-sized charcoal. 'Years later I was thinking about a way of turning carbon into charcoal,' he told *The Times*. 'So we came up with Carbonscape.'

'The problem with standard methods of pyrolysis is it's hard to control the type of material you get,' says Tim. 'Different raw materials create charcoal, gas and oil with different characteristics, and so do the conditions you burn it under, the mix of oxygen, the temperature and so on. To get a consistent result you've got to cook it right. That's where the Phantom comes in: consistent results.'

Those 'consistent results' make Carbonscape of huge interest to many biochar advocates and are the reason I'm here.

Certain types of charcoal added to soil can radically improve agricultural yields. In 2009, Dr Paul Blackwell of the Department of Agriculture and Food in Western Australia and colleagues reviewed soil charcoal studies from around the globe and were able to quote regular crop yield improvements of over thirty per cent and as high as fifty per cent. But they were careful to point out that 'variability is high and it is not yet clear under what soil and climatic conditions and plant species high or low yields can be expected.' The right mix can send yields rocketing, the wrong type can reduce them. A study by Imperial College London found that using soil charcoal improved UK barley yields substantially but *only* when large amounts of artificial fertiliser were also applied. For soils without added fertiliser, increasing the amount of charcoal slightly *reduced* the crop yield.

Part of the problem is that no one is quite sure why charcoal tends to have a beneficial effect on soil fertility. The general idea is that it is more effective at retaining nutrients in the soil (and keeping them available to plants) than other organic matter.

'The more we understand the relationship between charcoal and soils, the more we hope to give farmers or foresters the ability to "dial up" exactly the right charcoal to increase their yields,' says Tim. The advantage for the planet is that this will not only increase soil fertility, it will take carbon out of circulation. It's a possible win-win scenario.

'Actually, it's a potential win-win-win situation,' says Vicki. 'There's the oil and gas they can get out of the process, creating fuels out of waste.'

The development of the Phantom has taken a long time and seen a few hair-raising incidents. I'm not surprised. After all, it is basically a big experimental cooking pot with a massive industrial microwave attached to it. 'During one of the early experiments the cauldron's lid blew clean off,' says Greg the blue-suited engineer as he climbs down from the pot.

Along with his colleague Forrest, he shows me the Phantom set-up. Along the back wall of the shed in which the machine lives is a bench strewn with apparatus and tools. At one end is an ordinary kitchen microwave bearing the brandname 'Zip.' Propped up at a jaunty angle with wires and sensors coming out of it at various angles, it looks like it's come out of a firefight. Several blackened areas of plastic bear witness to its rude treatment. 'We test out basic ideas on this,' says Forrest. There's a kind of childlike glee in his eye that I'd also seen when Greg had mentioned the lid-blowing incident. I get the impression he might have been a kid who liked to set fire to things.

'Does the Phantom use a lot of power?' I ask. Normal pyrolysis might not give you much control but it is cheap; the biomass that becomes charcoal is the only fuel used. By contrast, this beast has large industrial power cables attached. I'm worried it might generate more carbon dioxide than it locks away.

'It does use power, and that's the price we pay for control. But we'll lock away twice as much carbon as we use. We're also looking into using the oil and gas produced by the pyrolisation to power it,' says Tim. 'The idea is it becomes self-sufficient.'

Later that afternoon Tim, Vicki and I take a trip to Picton, where Tim has his yacht *Faith* moored. It's another business venture as well as a personal pleasure – he hires the boat to groups of tourists for dolphin watching. The boat has some history too; it was once owned by Lord Shawcross, a British attorney general and chief prosecutor at the Nuremberg war trials that followed World War II. Shawcross successfully argued that 'just following orders' wasn't an admissible defence when it came to collusion in torture, murder and genocide, paving the way for the new era of soldiers' individual accountability for their actions.

Tim takes us on a boat trip into Queen Charlotte Sound. As we leave the harbour, huge hills rise on either side of the water, magnificent and imposing against the deep blue sky. It's a glorious sight – nature's raw beauty painted on land and sea. The land to the left and right of us is covered with trees, and the sight of all this 'biomass' inspires me to raise one of the objections to the proposed charcoal renaissance.

Some people worry that biochar might share a similar trajectory to crop-based biofuels, misappropriating land and taking it away from food production or disadvantaged peoples. In a largely out-of-character rant in the *Guardian* newspaper in 2009, the environmentalist George Monbiot characterised the idea as 'woodchips with everything' and said it amounted to turning the planet's surface into charcoal. '[The] proposal boils down to this: we must destroy the biosphere in order to save it,' he wrote.

Monbiot is worried that when biochar activists talk of planting trees to turn into charcoal, they often refer to growing those trees on 'degraded land' and that this is simply 'the new code for natural habitat someone wants to destroy,' land occupied by 'poorly defended people whose rights and title can be disregarded.' Advocates of biochar like James Lovelock, NASA scientist Jim Hansen, Chris Goodall and Australian global warming activist Professor Tim Flannery (who is on the board of Carbonscape) 'should know better' wrote Monbiot. (Just to put that in context, that's like someone telling Jimmy Page, John Lee Hooker *and* Andrés Segovia that they're not proper guitarists.) In fairness, Monbiot was a bit fed up with biochar hype and let rip. It's been heralded as a panacea, a complete answer to the global warming crisis – the cheap, easy solution. In reality, it's just one tool in the bag (although potentially a very powerful one).

Another concern is that as legislation comes into play insisting that more and more organisations offset their carbon emissions, they might turn to the cheapest solution: biochar, which, ironically, might lead to widespread deforestation. When eleven African countries asked the UN to consider international regulations to allow nations and companies to offset emissions using charcoal, there was outcry from some quarters. Rachel Smolker, a biologist and anti-biochar activist gathered signatures from nearly a hundred and fifty concerned organisations to protest against the adoption of what she called a 'charred earth policy.' 'It would require huge areas of land to be turned into plantations,' warned Smolker.

'Well, I don't think anyone's suggesting deforestation,' says Vicki, slightly bemused. 'It makes most sense to use agricultural and forest waste. I mean, there's so much of it, so many types around the world. Still, if the legislation is right, it could stimulate responsible reforestation.'

'Planting trees in order to cook them?' I ask.

'Yes. That's the whole point. You plant fast-growing species, pyrolyze them and then replant. The amount of forest on the planet could end up being much higher than it is now. But Carbonscape's business is about maximizing what you can get out of *waste* biomass.'

'What you can get out of waste biomass' turns out to be more than charcoal for soils, oil and wood gas. Tim talks about producing high-value chemicals that can be extracted and refined to provide further revenue for farmers. (He won't tell me which chemicals because the patents on the process are still being filed.) Vicki hopes that the control the Carbonscape process offers will soon yield 'activated carbon.' Activated carbon is an extremely porous form of charcoal that, because it has so many holes in it, presents a huge surface area to the outside world relative to its size. This gives it manifold uses – from absorbing pollutants, to air purification, to water filtration and temporarily neutralising the effects of poisons. You can even buy 'flatulence filtering underwear' that passes your farts through an activated carbon filter to neutralise the whiff. Yes. Really.

Half an hour after leaving the harbour, we drop anchor at Kumototo Bay, possibly one of the most tranquil places on the planet. It is startlingly quiet, the water barely moves. Only the occasional flapping of wings as a sea bird passes overhead disturbs the peace. After the whirlwind of new knowledge I've been stuffing into my brain on this journey, it's a welcome moment of calm. I can almost feel my brain slowing down a few clicks.

Vicki has supplied freshly cooked lobster that we eat with our fingers while sipping a locally-made Chardonnay brought from below decks by Tim. I feel incredibly lucky to be here. Vicki and Tim are an ideal mix of easy company and sharp intellect. Conversation and ideas flow easily, like bubbles coming to the surface of a fizzy drink. Everything about this country and these people seems like a release.

Tim tells the story of an elderly German businessman and his wife who had hired the boat to go dolphin watching. It was one of those days where the creatures just weren't coming out to play, but the businessman was dogged in his determination to see them, requesting Tim to try ever more unpromising spots when the usual viewing places had proved empty of our aquatic cousins. After about two hours, Tim anchored the boat for a drink and a snack in a quiet spot like the one we're now sitting in. 'Do you want to continue looking?' he asked his guests. 'Or would you just like to sit and be?'

'The funniest thing happened,' says Tim. 'He looked at me as if I'd suggested the most radical thing ever. "Sit and *be*!" he said. "Sit and *be*!" – as if it was the first time the idea had ever occurred to him. He called to his wife. "Write this down!" he told her. "Sit and be!" "Do you want to keep looking for dolphins?" I asked and he said "No! I want to sit and be!" so that's what we did.'

We're all laughing, just sitting and being. The three hours I spend on the boat are some of the best of my entire trip.

That evening we have dinner in *Le Café* on the seafront at Picton where I'm introduced to another co-director of Carbonscape, Nick

Gerritsen, his gallery-owning partner Barb Speedy and a man called Don Binney who, I'm informed, is one of New Zealand's finest artists, famed for his celebratory depictions of birds.

Nick Gerritsen is a firebrand, almost aggressive. Physically he looks like a cross between a young Gene Hackman and John Lennon, circular specs sitting over eyes that give the impression of being searchlights seeking out opportunity. Or weakness. Nick is a 'knowledge broker' – a man who puts people, ideas and money together. He's clearly not someone to suffer fools gladly. In fact, I get the impression he regards people who don't come up to his intellectual mark as walking versions of waste biomass. He's just come back from Australia where he's reached the end of his tether with the nation's use of the word 'mate' to refer to just about anyone. He's in a combative mood and wants to know whether my book is justifiable. 'What kind of an optimist are you?' he asks on hearing the title.

'Yes, I hope you're not a *Panglossian* optimist,' says Don Binney, drawing a blank expression from me in return. It turns out he's referring to Voltaire's satirical book, *Candide*, in which the character Pangloss, who sees all history through rose-tinted spectacles, is used as a device to lampoon the optimistic philosophies of Leibniz.

Bloody hell. After the relaxed conversation on the boat, it's suddenly got all gladiatorial. I've got a famous painter quoting French philosophy at me on one side and an *agent provocateur* fresh from Aussie-bating on the other.

'Well, actually the book didn't start off with the word "optimist" in the title,' I say. 'I just wanted to find out what was coming next. But when I told my agent Charlie about some of the stuff I was investigating he said, "this sounds great, this is optimistic stuff" and I realised he was right. There are people just getting on with incredible things that we don't really hear about enough. He put the "optimist" in the title, not me. So, I'm a *convinced* optimist.'

Don Binney seems happy with this. Nick considers for a moment. I decide to take a risk. 'You seem like you need to let off steam,' I say to Nick. 'Like you've got Argument Tourette's.'

Vicki laughs uproariously and the mood lightens. Nick lets go of his Australian frustrations and our conversation instantly becomes more like a good game of cards instead of the fencing match I'd feared. It seems no one in New Zealand takes themselves that seriously for long. I use the opportunity to ask Nick a few more questions about Carbonscape.

'Did you name The Phantom?'

'Yes. The engineering team wanted something more credible sounding but I figure people have long respected and loved the individual personality of machines.'

'It's worked in our favour,' says Tim. 'It quickens bored journalists into interest. Let's face it, charcoal is pretty dull.' True to form, the company's coming machines have equally comic book names.

'The Phantom is really our research test bed,' says Nick. 'We're now commissioning a machine that can process biomass continuously.'

'What's that called?' I ask.

'The Rototron.'

'That's actually based on the term "magnetron", the device that generates microwaves,' says Tim. 'So it makes sense.'

'We're also working on a portable machine for rapid analysis of biomass so we know what is possible to get out of it in terms of oil, gas, biochar and other products,' continues Nick. 'This will help quickly work out what you've got and select the cooking program to get the outcomes you want.'

'I'm scared to ask . . .'

'It's called Zippy Bling,' says Tim, smiling. 'I named that! I was inspired by our bashed-up Zip microwave. The name, like the microwave, has held out against the odds.'

Despite the jollity and comic book naming of machines, I'm wondering if I'm seeing the birth of a new industry here. Taking waste and turning it into useful products isn't a new idea but suddenly I'm seeing it writ large. With billions of tons of agricultural biomass essentially going to waste, there is an enormous opportunity. If Carbonscape or one of their competitors can help turn that into useful products and ways for farms and forests to earn extra income, we could see a rapid uptake of charcoal making, taking carbon out of the carbon cycle and realizing Lovelock's 'one way we could save ourselves.' Carbonscape wants to sell Phantoms and Zippy Blings and Rototrons to anyone who's got waste biomass.

'We want to get the technology as close to the end user as possible,' says Nick. 'The way I see it, this is a mining company. Except we're mining waste biomass off the surface of the Earth rather than ancient biomass from under it.'

'Renewable mining?'

'Something like that, yeah.'

Someone around the table suggests that there should be a cocktail based on the title of the book, and much laughter ensues as we try to decide what should go in it and what it should be called. Nick calls over to Peter Schöni, the café's owner and barman, and we task him with creating 'An Optimism.'

'What do you want in it?' asks Peter, turning to me.

'Well, we've decided it needs tequila and champagne, but the rest is up to you,' I respond.

Twenty minutes later, after some experiment, Peter returns with a champagne flute filled with a red drink.

'What's in it?' I ask.

'Take a sip,' he says.

It's delicious; the perfect drink for a summer evening. I write the recipe down on a napkin which will end up framed in my flat: a shot of triple sec, a shot of tequila, lemon juice and strawberry pulp, shaken over ice and topped up with champagne. We collectively decide the drink shall be known as either 'An Optimism' or 'An Optimist' except here in *Le Café* in Picton, where it will have the special name of 'A Schöni Optimist.'

If you're ever there, please try one and tell Peter I sent you.

CHAPTER 12

A little bit of a bloody big amount

The nation that destroys its soil destroys itself.
FRANKLIN D. ROOSEVELT

It is the mid-1960s. Zimbabwe is in the grip of a civil war which has its roots in white colonialism. As part of the war effort, a young scientist is put in charge of a 'tracker combat unit.' Even though he is hunting humans he is constantly observing and thinking about the condition of the land over which he pursues his quarry. He tracks people over game areas, tribal areas and commercial farms. Everywhere he inspects plants and soils, looking for the signs of human passage. What he learns may one day help provide an answer to man-made climate change and improve the prosperity of nations.

A shift of time and continents: to 2009, Australia. Michael and Anna Coughlan have just bought a farm, Moombril near Holbrook, New South Wales, and are about to surprise their new neighbours by selling off a large number of its assets. These include two houses, the shearing shed, the hay shed, the machinery shed, two garages (and the four vehicles in them), even some spare sheep dip. At the sale, whispers of incredulity can be heard passing among prospective local buyers.

Using part of the money raised, they purchase some fencing materials and two motorbikes. People think they are crazy.

It's about a year after the Coughlan's farm sale and I'm in New South Wales myself, flying in to the tiny airport of Armidale with a farmer called Bruce Ward. We are met there by his business partner, Tony Lovell.

Tony is a big chap, tall, solid and with a manner you can only describe as 'cheeky.' Bruce is a bit more serious. Tony starts sentences with 'Imagine if . . .' whereas Bruce will open with 'The thing you need to realise is . . .' Tony will laugh out loud, Bruce will chuckle knowingly. But Tony's playful nature can't hide his serious intent. He and Bruce are going to save the planet.

As Bruce and I load our bags into their Landcruiser, Tony tells me that our first port of call will be a farm that sacrifices Poms in an ancient soil fertility rite. I am to be an offering to the gods. This sets the tone for the next few days. We will cover over a thousand miles as Tony and Bruce take me to a string of farms that are redefining the way we think about agriculture and that offer a model for mitigating global warming while delivering sustainable increases to food production and bringing the cost of that food down. What's so striking about what they propose is that rather than using new technologies these farms are using a very old one. 'The thing you need to realise is it's a technology that's been around for millions of years,' Bruce tells me.

In the process I will learn what links a tracker in the Zimbabwean civil war to Michael and Anna Coughlan's asset sale in modern day Holbrook.

Australian farming has become synonymous with drought. A decade of low rainfall, heat waves and wildfires has scorched much of the nation. Australians call it 'The Big Dry' and it's brought acute hardship to rural communities. When the rains come, as they did to some parts of the country in early 2010, water runs over the parched surface often resulting in devastating floods. It's as if the land is now so unfamiliar with water it no longer knows how to drink. The drought, it is said, remains deep in the soil. Farmers across the continent are suffering. Agricultural debt has increased from just over ten billion dollars in 1994 to nearly sixty billion dollars in 2009. Many of the farms we drive past on our journey are surviving on 'drought assistance' payments handed out by the government. Australia is worried about food production. Some believe the only way for the country to survive is with a reduced population. I'm here to investigate another solution.

'Imagine if you could take billions of tonnes of carbon dioxide from the atmosphere every year, safely, effectively, economically and immediately,' says Tony as we get underway. 'Imagine if you could do it in a way that also increases biodiversity, boosts food security, reverses the advance of the desert, and improves rural communities.'

I'd first heard this pitch in Manchester, England, before setting out on my journey. Tony had talked at an event called 'The Manchester Report' held in the gothic splendour of Manchester Town Hall, a kind of serious game show for climate change initiatives. Over two days, twenty short-listed ideas were presented to an expert panel and live audience. Tony's presentation stood out for two reasons. First, he's a very engaging speaker, combining authority with an approachable and light manner that makes you want to go to the pub with him. Second, what he said was extraordinary. The panel was 'hugely impressed' and so was I. It seemed too good to be true.

I introduced myself to Tony after the event. Bruce was there too, but couldn't get a word in edgeways as Tony and I fell into the worst kind of joke one-upmanship, something we've slipped right back into here in Australia. During long hours travelling, Bruce sits in the backseat alternately chuckling and dispensing wisdom against a barrage of puns and innuendos. When Bruce suggests a turning to Eubalong (pronouncing it 'You-abbah-long), for instance, Tony replies with 'You have a long what? Try to keep it clean, Bruce.' Everything they say about Australian humour is true. This goes on solidly for four days.

The jaw-dropping moment in Tony's Manchester presentation involved two photographs. 'This is a typical ranch in Mexico,' he explained, showing an image of an arid terracotta dustbowl with sparse vegetation and bare, compacted soil. Then he put up a second image showing a property awash with lush green vegetation. The contrast couldn't have been stronger, which made what Tony said next so astonishing. 'This is the ranch next door. Same soil, same rainfall. These pictures were taken on the same day.'

Tony and Bruce have a bunch of these photos, which they share with me over the coming days. Some of the most striking are aerial shots that show neighbouring properties from above. The difference between vegetation and bare soil follows the line of the fences separating adjoining farms.

'What's the difference?' I ask.

'Management,' says Bruce. 'Just management.'

About an hour after leaving the airport we arrive at Lana, the property of Tim and Karen Wright, where they farm Merino sheep. Tim isn't convinced about man-made global warming. He thinks politics 'has got in the road' and 'there's a lot of hidden agendas going on at the IPCC.' That hasn't stopped him and Karen from following Bruce and Tony's advice. We sit drinking lemonade on the veranda overlooking a green and fertile landscape. It looks more like Sussex than the images of the Australian bush I'm used to seeing on TV.

Tim and Karen ask me where I've been on my travels. When I mention my trip to Konarka, Karen disappears inside. A moment later she brings out a sample of the company's Power Plastic, the same stuff I'd seen rolling off that 'big printer' in New Bedford. 'We're a distributor for New South Wales,' she says.

I guess in a country synonymous with drought it makes business sense for a farmer to diversify into selling solar panels.

'Is this a business for if the drought gets worse?' I ask.

There's an awkward silence. Clearly I haven't got it.

'We don't feel the drought like other people do,' says Tim. 'Our reservoirs are still three-quarters full.'

The Big Dry isn't so big here it seems. But though Lana isn't feeling it, many of its neighbours are. Local farms that have been in families for generations are being sold off as the farmers have become unable to make a living. 'Some are suicidal,' says Tim grimly. He should know, having worked as a counsellor. His talk of rural suicide won't come as a surprise to anyone who lives in Australia. A widely quoted statistic is that one farmer kills himself every four days. That figure is actually a decade old but things have hardly improved in the intervening period.

Tim tells me the rainfall has been low for *nine years*. Yet despite this, Lana is flourishing with high levels of livestock, tripling since 1980 from 7,000 to 22,000 'dry sheep equivalents,' or DSE. When I started this trip, I had no idea I would find myself getting to grips with sheep-based metrics. DSE figures are a way to compare farm capacities and performance. 'One DSE' is the amount of feed needed by a two-year-old Merino sheep of the sort Tim and Karen farm. A pregnant ewe might be 1.5 DSE, a cow nursing a calf ten times that. Bruce tells me the usual labour ratio is one man per eight thousand DSE. Tim is running at one man to 15,000.

'Tim and Karen are operating at full capacity with months of feed in front of them, where all around them is panic,' says Bruce. 'The neighbours look at us with envy,' adds Karen. 'They say "you must have had a lot of rain here."' She laughs, though without humour. 'We've had the same rain as everyone else.'

Something is clearly going very right with this farm, but it's not the only one – and Tony is eager we get on the road so I can see more. We've a long way to go.

We've been driving for about an hour when Bruce asks Tony to pull over. 'Let me show you something,' he says.

As we get out of the car Bruce directs my attention to a paddock by the side of the road. The ground is almost totally bereft of vegetation – the sort of landscape you expect a man in a spacesuit to be walking across.

'Is that the lack of rain?' I ask.

'Look down the fence,' says Tony.

And then I see it, and from that moment on, I see it everywhere we go. Outside the fence there's grass. In fact, we're standing in it up to our knees. Inside the fence it's practically desert.

'Things can grow just fine here,' says Bruce. He bends down and grabs a lump of vegetation. It comes away in his hand. 'This is dying, though,' he says. 'That's the other half of the problem.'

Back in the car Bruce tells me the story of a man called Allan Savory, who pioneered the farming methods employed by Tim and Karen and everyone else I'm going to see on this journey into Australia. Savory was the previously-mentioned tracker in the Zimbabwean civil war of the 1960s, and before that was the provincial game officer in the Northern and Luapula Provinces of Zambia (then the protectorate of Northern Rhodesia).

'Savory was trying to understand the continuing problem of land degradation, like what you've just seen in that paddock. His first instinct was that it must be something to do with the cattle, things that humans had inserted into the landscape in place of the natural wildlife. But it turned out it wasn't the animals, it was the way they were being managed.'

Tony interjects, 'If you go back in time, our grasslands were dominated by large herds of grazing animals, bison in America and wildebeest in Africa.'

There are still a few places where you can see what Tony is talking about. The Serengeti, for instance, is one of the few remaining natural grasslands on the planet and is home to huge migrating herds of wildebeest and zebra. There is no beginning or end to their journey, but a constant clockwise pilgrimage in search of water and fresh grass. Every year roughly two million animals cover eighteen hundred miles. The herds stay closely packed as a defence against predators like lion, cheetah and hyena.

'What happens is the herd eats the grass but then moves on looking for the fresh stuff. In the Serengeti, that herd won't be back on the same ground for up to a year or so,' says Tony.

'That's important,' says Bruce. 'Savory realised there is a natural relationship between grasses and grazing animals. The growing buds are at the base of the plant and they need sunlight. If the plant gets too tall it starts to kill itself by hiding those buds in its own shade. It can't photosynthesise.'

'That's why that grass came away in my hand,' explains Tony. 'It's dead material. In nature the herd would have come along, eaten the tops off the plants exposing the growth buds and moved on. By the time they came back, the grass would have regrown. The animals have another go and so on and so on. The plant *needs* the animal to keep it alive.'

'The problem with the way we farm livestock is we don't let them roam,' says Bruce. 'We split up big herds between separate paddocks and keep them there for way too long. With no predators they can wander where they like in that space. They don't stick together, which means they quickly eat all the grass in an area. Worse than that, the grass never gets a chance to grow back. An animal will have a go at it as soon as it starts sprouting. It's a vicious circle, with the animals destroying their own food. That's what you saw in that paddock back there: desert inside the fence and tall but unhealthy plants outside it – overgrazing and undergrazing side by side.'

For the next four days I see this pattern of overgrazing and undergrazing repeated everywhere we go: miles upon miles of fences separate sparsely vegetated paddocks from verges of dried-out and withering foliage.

'Why doesn't anybody notice this?' I ask.

'Did you notice it?' responds Tony. It's a fair point.

Bruce, ever wise, says, 'When you've grown up with something that's always looked one way, then you can't see it should look any different.'

We spend the evening at the Overlander Motel in Gunnedah, the 'Koala Capital of the World.' Checking in I notice the establishment sells a product that only Australia could have come up with – the 'StubbyGlove.' Dubbed 'the greatest innovation in beer drinking technology since bubbles,' it's a glove made of neoprene rubber with an inbuilt beer bottle holder. ('Stubby' is Australian slang for beer bottle.) The idea, say the inventors, is that if you fall down you won't drop your beer. I like the idea of a product that allows you to hold on to precious things during accidents, but only in Australia is falling over more of a trial if you've lost your *beer* and not just your balance.*

'Is that for real?' I ask the receptionist, finding it hard to hide my amusement.

She ignores me and turns to Tony.

'What's his problem?'

'He's a Pom.'

'Ah,' she says nodding. Clearly this explains everything.

We eat in the motel restaurant, which serves up simple hearty meals heavy on meat. I use the opportunity to talk about the main reason I'm here: the climate change angle of Tony and Bruce's work.

'Look,' says Tony. 'Even if there was no man-made global warming problem, what Tim and Karen are doing makes sense for their business. The good news is that the environmental benefits come screaming out the back end of it anyway.'

Those 'environmental benefits' include the removal of carbon dioxide from the atmosphere at potentially massive levels as well as increasing biodiversity. Tony elaborates: 'There has been a huge decline in soil carbon levels across the grasslands of the world over the last hundred and fifty years which is directly related to the loss of vegetation. That paddock we stopped at? The soil carbon there would have been almost nonexistent.'

* The product originally had slightly loftier ambitions. The idea was born during a skiing trip to the mountains of New South Wales. Glen Krummel and his brother Leon watched their mate Jim, who had lost his arm in an accident years previously, struggle to keep his hand warm but his beer cold. The StubbyGlove was born, and as soon as the friends started using them, publicans began asking for branded versions. They've sold over a hundred thousand.

Beers arrive and we elect to use our own hands to hold them despite the StubbyGlove option. As we drink, Bruce explains that grass plants grow roughly the same amount of root matter as leaf matter – they're in balance above and below the ground. If the plant gets nibbled by a cow or sheep, it'll slough off a corresponding amount of root matter into the soil within minutes.

'Now a plant is fifty-eight per cent carbon . . .' says Bruce.

'. . . and carbon is good for the soil,' I say, remembering what I'd learned on my trip to New Zealand.

'Very good,' says Tony. 'Clearly schooling has improved in the UK.'

'But that carbon can't do anything to offset global warming, can it? I mean, nearly all of it is going to return to the atmosphere as the biomass rots.'

'Right again,' says Tony. 'But the crucial words in your last sentence were "*nearly* all." When organic material decays, it leaves behind a small amount of residue in the soil.'

'Humus,' says Bruce, giving that residue its scientific name.

'No thanks,' says Tony. 'I think I ordered the steak.'

Bruce ignores this (he's used to it). 'That residue, the humus and charcoal components stays in the soil in a natural system.'

'It's only a small amount of the carbon that made up the plant,' says Tony, 'but a little bit of a bloody big amount soon adds up.'

'Does this work in reverse?' I ask. 'If there isn't vegetation, does the soil start to lose that stable carbon?'

'Yep. That's why conventional agriculture can bugger up a pristine ecosystem so quickly. That's what we've been driving through all day.'

I'm arrested by a thought and look Tony straight in the eye.

'So let me get this right. You're saying agriculture has released billions of tons of carbon into the atmosphere *from the soil?*'

'Yep.'

'You're telling me that the culprit in global warming isn't just the Industrial Revolution, but the *agricultural* one?'

Tony looks over at Bruce. 'He picks things up quick for an English fella, don't he?'

That's probably because I'm remembering something I read about the work of paleoclimatologist William Ruddiman, whose book *Plows, Plagues and Petroleum: How Humans Took Control of Climate* argues just this point. In fact, according to Ruddiman, agriculture is the reason we're not in an ice age right now.

Ruddiman's book reminds us that the planet is subject to a dance of astronomical cycles that affect how much of the sun's radiation reaches us. For the best part of the last million years, this means the Earth has experienced regular glacial cycles, with ice covering about a quarter of the planet's surface for a hundred thousand years at a time, interspersed with short ten-thousand-year 'interglacial' warm periods. The last warm period should have ended two thousand years ago says Ruddiman, but agriculture has warmed up the atmosphere and held it off. 'Farming is not nature,' he writes, 'but rather the largest alteration of the Earth's surface that humans have yet achieved.'

According to the UN's 'Livestock's Long Shadow' report, the total area of land given over to grazing amounts to twenty-six per cent of the ice-free surface of the planet. If, as Bruce and Tony are telling me, we've been managing that land in such a way that our soils have been losing carbon, then as agriculture has grown, so has atmospheric CO_2 as a result. This is not to underplay the crucial role of fossil fuels in accelerating that rise, but it is wrong to lay all the blame at their feet.

Some commentators worry that cattle burping methane (another potent greenhouse gas) into the sky is a major challenge to the climate and call for us to relinquish meat eating. Methane is indeed twenty-three times more potent than CO_2 as a greenhouse gas and the UN estimates that livestock contribute eighty-six million tonnes of it to the atmosphere every year.

But if what Bruce and Tony are saying is right, getting rid of cattle could end up being a disaster. It's not that we have too many cattle, it's that we have too little grass. The soil needs the grass if it is to become richer in humus and therefore carbon. The grass needs animals if it's not to die from undergrazing.

Working correctly together, the system is a huge natural 'carbon pump' that can take massive quantities of CO_2 from the sky.

'The good news is that we only need to change the way the cattle move to make things better,' says Bruce. 'It's something we can do today and it more than pays for itself. The land is the biggest thing we can control. We can't manage the seas and we can't manage the air. But we can actively manage the land. It's the single biggest asset we have in battling climate change and building biodiversity.'

Tomorrow I'll find out just how big the impact could be.

After an early rise and a six-hour drive taking us three hundred miles into ever more parched country we've come deep into an area known as the Western Division. Our journey had one stop – breakfast at the Jolly Cauli in Coonabarabran, a café that bizarrely doubles as an osteopath. This mystifies me until I see Tony order his breakfast, which seems to contain the meat of an entire cow. 'My grandma said to me, you don't get to this size by being fussy!' My guess is the osteopath makes a good business dealing with back injuries caused by standing up too soon after eating here.

Now we've arrived at Etiwanda Station, a farm operated by Andrew and Megan Mosely in Cobar, one of the driest places in New South Wales. Soon after arriving, Megan says, 'When the TV wants a few shots to represent drought they film out here.'

It hasn't rained in Cobar for five and half months but the Moselys are calm. Their farm is a beacon of prosperity in one of the most stressed parts of the country.

'The land hit its low point in the late 1990s,' says Andrew, handing out mugs of tea. 'Back then we'd need fifteen acres to keep one sheep alive.' In his eyes you can see the memory of bad times. 'It got so degraded it was never going to recover under normal management,' adds Megan.

'You were in a pretty dire state?' I ask.

'Yeah, we were not exactly sitting pretty. To be honest we were desperate. We needed to make some smart choices and in the end we had to completely change the way we managed the land.'

'It was either that or give up and walk away,' says Andrew.

'Is it the same for your neighbours?'

'Many of them are living on government support. They're existing but they haven't got any kind of lifestyle. Their houses are falling down.'

Megan and Andrew's house isn't falling down. In fact, it's a comfortable, well-furnished home with modern conveniences. Sprinklers keep the lawn outside a lush bright green, and healthy-looking pets lounge in the blistering heat, occasionally teased by the couple's two young daughters, Jessica and Emily. There's a small pig house near the property and I ask the girls the names of the two pigs who live there.

'Easter and Christmas,' says Jessica.

'Those are nice names,' I say, before realizing what they mean.

Andrew takes us on a tour of the property.

'We're seeing native perennial grasses returning. We're not sowing any grass seeds, they're just coming back naturally,' he tells me.

'Where's the livestock?' I ask.

I haven't yet seen a sheep. And for a sheep farming operation, this strikes me as odd. By way of explanation Tony, who is doing his own bit to irrigate the land, says, 'One of Allan Savory's central ideas is you get all your animals into big herds, like they would be in nature.' Zipping up his fly, he continues, 'What that means is that most of the property won't have animals on it for most of the time. They're all together in a small area.'

'That's what gives the grass time to grow back,' says Bruce.

'How do you keep them all together?' I ask.

'That's easy. You build lots of small paddocks. Leave them in each one for a day or two and then move them to the next.'

'The biggest cost is fencing,' says Andrew. 'You have to put in a lot of fences to divide up the property that way.'

After all my investigations into the cutting-edge of technology and medicine I'm struck forcibly by what Andrew has just said.

'One of the answers to global warming is *fencing*?' I ask. 'This is the technology we're talking about here? You're saying pulling carbon dioxide out of the sky is a matter of fences?!'

'Yeah,' says Tony, shrugging. 'We reckon it's something even a Pom can get his head around.'

Over dinner I turn the conversation to soil carbon levels. I want to know if the Moselys have seen levels rise the way Bruce and Tony are saying they should.

'We had some scientists up here from Trangie Research Station and the University of New England,' Andrew tells me. 'They reckon the carbon went up about fifty per cent over three years.'

'It was ridiculously low to start with, though,' Megan interjects. 'So fifty per cent of very little is still very little.'

Tony wants to know the exact figures. Megan pulls out a file and reads carefully.

'It went from 0.6 per cent to 0.88 per cent in the top fifteen centimetres of the soil,' she says, her finger tapping the page. 'It'd be higher now. We've had a bumper grass crop this year.'

Lightning fast, Tony does a rough sum in his head. 'That's about five tonnes of carbon per hectare then,' he says.

That sounds like a lot of carbon for such a small percentage change but the figures stack up. An average soil weighs about 1.2 tonnes per cubic metre. A hectare is ten thousand square metres so digging down a metre gives you twelve thousand tonnes of soil. An increase of just 0.28 per cent in soil carbon as seen at Etiwanda Station therefore amounts to nearly thirty-four tonnes per hectare measured to a metre deep. Andrew and Megan's soil had only seen new humus penetrate the top fifteen centimetres. In time root matter will easily reach further, but for now it means that figure of thirty-four tonnes is reduced to a smidgen over five tonnes per hectare for the Moselys' farm. Tony was spot on. That's fast maths.

In relation to climate change, though, there's another step in the calculation. Every extra atom of carbon that's now in the soil has got there thanks to photosynthesizing plants, which means it was originally part of a carbon dioxide molecule in the atmosphere. Because carbon accounts for only 27.3 per cent of the weight of a carbon dioxide molecule, five tonnes of soil carbon is the equivalent of roughly 18.5 tonnes of CO_2 pulled out of the sky. Extrapolating the increase in soil carbon over the Moselys' 26,000 hectares gives a figure of 480,000 tonnes of sequestered CO_2 from the atmosphere.

To put this in context, this is equivalent to offsetting the CO_2 emissions generated by over eighty-five hundred Australian citizens over the three-year period (and when it comes to per capita emissions, Australia is one of the world's worst offenders, with figures nearly triple that of your average European). As Tony said in the restaurant last night, 'A little bit of a bloody big amount soon adds up.'

'For pretty much any soil type, if you increase organic matter by a full one per cent to a depth of thirty centimetres, you sequester roughly a hundred tonnes of CO_2 per hectare,' says Tony, taking a suck on a stubby. 'It doesn't matter if it's degraded soils like here and you move from say 0.5 per cent to 1.5 per cent, or a Scottish moor where you've got fifteen per cent organic matter already in the ground and you take it to sixteen per cent.'

Published research is full of examples showing how changes in grazing management can deliver increased levels of soil carbon. For instance, the Blackstone Ranch in Ranchos de Taos, New Mexico, sequestered over 13,600 tonnes of CO_2 over ninety-four hectares in just one year (that's nearly 145 tonnes of CO_2 per hectare per year), while doubling its livestock and increasing gross income by 395 per cent.

The UN's Food and Agriculture Organisation estimates there are 3.5 billion hectares of agricultural grasslands on our planet. If we could increase organic matter in these by one per cent, this would offset nearly twelve years of global CO_2 emissions. Some nations have already caught on. In July 2009, the Portuguese government introduced a multimillion-dollar soil carbon offset scheme based on dryland pasture improvement covering forty-two thousand hectares and compliant with the Kyoto

Protocol. The scheme pays farmers to establish biodiverse pastures to increase soil carbon.

Paying farmers for soil carbon increases could encourage more landowners to follow Andrew and Megan's example, and this is something Bruce and Tony have been tirelessly campaigning for.

After dinner, I ask a question that's been in the back of my mind since yesterday and one I'd nearly put to Tim and Karen.

'Could you not take over failing farms and do to them what you've done here? Are you tempted to buy other properties?'

Megan laughs. 'If I had a bucket of money I'd buy the whole Western Division. You could do it; do up each farm, put in the fences, get the stock moving right. Yeah, you could take on places and restore them like you restore a house.'

'If you had a bucket of money?'

'If we had a bucket of money.'

Tony is smiling and I think I know why. There's a reason he can do complicated arithmetic in his head in double-quick time. You see, he has another side to him. He isn't a farmer, or an agricultural scientist, or a soil carbon expert by profession. He's an accountant, and a very successful one.

A new day, another six hours on the road and a further three hundred miles bring us to the house of Ken Haylan, a retired businessman who now farms cattle at Hailiebrae outside the town of Blayney. Ken is in his sixties and manages the farm pretty much by himself. He's been following Bruce's advice for two years and on the journey down, Tony had told me another of his too-good-to-be-true statistics and I want to find out if he's pulling my leg.

Over lunch I ask Ken about his figures. 'Tony tells me your costs of production are pretty low.'

Ken considers for a moment. 'Yes. Although we're not fully stocked at the moment so it's a little higher than it could be.'

'Can I ask you your per-kilogram figure?'

Ken tells me, my jaw drops and Tony smiles. You see, the average cost of production for a kilogram of beef in New South Wales is Aus$1.60. Ken has just told me his costs are 'about fifty-three or fifty-four cents.'

He continues. 'At full stock it'd be about forty cents.'

'But that's a *quarter* of the average figure?! You're telling me you can sell your meat cheaper than your competitors can produce it and take a huge profit margin at the same time?'

'Something like that,' he says nonchalantly. 'Would you like another slice of quiche?'

I'm taking this in when Ken adds, 'What unites all the farms you're seeing is that they are 'low input' operations. They mimic ancient grazing patterns within the parameters of a modern agricultural holding. It is the antithesis of most modern agricultural practice with high cost inputs and reliance on a chemical fix for a problem rather than seeing the productive base, the land, as a whole.'

Since he's started following Bruce's advice, Ken says he has more grass in his paddocks when his cattle are leaving than other farms have waiting for their livestock entering a new field.

'Don't your neighbours wonder about that?'

'Oh yeah, but they'll never change.'

'The problem is that farming's a traditional business and it stays in the family,' Bruce explains. 'Changing the way you farm is the same as saying that what your pappy and your grandpappy did was wrong, which isn't easy when they're still living on the property.'

It's a common theme. Tim, Karen, Andrew and Megan had told similar stories. This is one of the reasons Tony wants Australia to follow Portugal's example and experiment with paying farmers for the carbon they sequester in the soil.

'It's a carrot instead of a stick,' he says. 'It's hard to change when everything's going to hell, when you've got to admit that the way you've been farming isn't working so good and all the while your dad's telling you to stick with tradition. But if you can pay them for the carbon, that gets them out of both holes – financial and emotional. The land improves, which is good for business, and they can tell pappy the change in management is to bring money in from something he never could have got paid for. It's not that the old man was wrong at all, it's just a new market: carbon.'

'But even without carbon credits, they can see the difference just looking over the fence, surely?' I'm slightly incredulous.

'Well, these are special Australian fences,' says Tony. 'They're impervious to new ideas.'

For once, he's not joking.

'Yep, you're never a prophet in your own dunghill,' suggests Bruce sagely.

One day later, two hundred and fifty miles south, and I'm in an argument with perhaps the smartest farmer in Australia.

Graham Strong is not short of opinions or knowledge. He's vehemently opposed to genetic modification, for instance. I've mentioned the argument that selective breeding of plants and animals is a form of genetic modification and am now listening to an extremely educated rant about the difference between transgenic modification of species (the practice of taking a gene from one organism and deliberately inserting it into the genome of another) and selective breeding. 'It's not the same and I hate it when people try to say they're equivalent!' Graham exclaims. I'd only raised it as a talking point, but our host has gone in for the kill. Tony is standing quietly to one side with the traces of a smirk on his lips. I think he's slightly enjoying the Pom getting an Outback-style bashing.

'Passionate' doesn't come close to describing Graham, who manages Arcadia at Boree Creek just outside the town of Narrandera. When he talks about the land, it's like he's referring to a brother, but don't let that give you the impression he's some tree-hugging hippie. During his polemic about genetic modification, he reveals an impressive level of knowledge about many of the subjects I'd discussed at Harvard Medical School.

He believes we need an urgent debate 'about how we operate in society, about the relationship between our urban centres and how we produce our food, share water, share resources.' Part of his contribution to that debate is getting involved in large public art projects that take an appreciation of the land to new audiences. 'You can't get upset with someone being ignorant about what you do if you've done nothing to tell them about it,' he declares. 'It's not that you can't farm this land. You just can't farm it the way you used to.'

Some people, though, may not want to hear Graham's most controversial view: he doesn't believe in drought.

With the hardships being suffered throughout the Australian farming industry, this sort of talk can get you lynched. 'I'm careful with the sensitivities, I understand them, but the word "drought" means nothing without a cultural context,' he tells me, his tone ever insistent.

'*Our* cultural context is the fact that our western lifestyle is at odds with the reality of our climate.'

'But you can't deny the rainfall is low.'

'Well, what's low? It's lower than we're used to, but there's still enough water if you manage the land right.'

In a repeat of what I'd seen at Lana, Etiwanda Station and Hailiebrae, a tour of Arcadia earlier in the day had shown me a healthy farm with abundant crops of saltbush, a feed for Graham's sheep which delivers leaner, more succulent lamb.

Part of the reason these farms are prospering, even during The Big Dry, is because when the rain does fall, their soils retain more water. According to the Soil Carbon Coalition, an Australian non-profit that works to spread soil carbon knowledge, 'organic matter can hold four times its weight in water.' This means that a one per cent increase in organic matter relates to an extra hundred and fifty tonnes of water storage available per hectare. For a nation in the grip of The Big Dry, that makes a huge difference and it's one of the reasons Tim and Karen were able to tell me 'our reservoirs are still three-quarters full.'

'The word "drought" implies impotence,' says Graham. 'That nothing can be done. End of conversation. Well, no thanks.'

Our last stop is a farm called Moombril near Holbrook, owned by Michael and Anna Coughlan, the couple who surprised their new neighbours by selling most of the property's assets as soon as they bought it. After another hundred miles and two hours of jokes, I'm finally looking at a cow. Having spent the last four days talking to livestock farmers and seeing only family pets and pigs named after the holidays they'll be eaten on, I've told my hosts I want to see some beef on the move so Bruce, Tony, Michael Coughlan and I are in one paddock, looking into another which contains two thousand cattle – one of the large herds that mimic those that roam natural grasslands.

These cows won't return to the paddock they're currently in for perhaps a hundred and fifty days, giving the grass they're now happily munching on plenty of time to recover. In fact, the paddock we're looking at them from (and where they'll head next) has grass up to our knees. As far as I can see, there's greenery.

Michael would normally ride out here on one of the two motorbikes he and Anna bought with the proceeds of their asset sale. 'It's just as easy to open a gate for two thousand cattle as two hundred,' he says.

'Is it really as simple as it looks?' I ask.

'Well, in principle, yes. The key to what we're doing is we've really dumbed it down, just simplified it and kept discipline with that. Because the animals are out here doing the work, the system kind of runs itself. We don't need to spend money and time on other inputs like fertilisers or pesticide sprays.'

Ken Haylan, the man who'd blown my mind with his tiny production costs, had said something similar. 'If you do it right, there's not much you should be spending money on,' he'd told me. 'You don't fertilise so you don't need a tractor; you don't need to spray because you don't have a major weed problem; and your cattle's health improves too.'

Neither Ken nor Michael need to drench their cattle for worms. (Ken drenches cattle he buys once on purchase, 'and that's only because I don't know where they've been.')

'It's tiring just watching some of my neighbours,' says Michael. 'I have been asked what I do all day.'

'What do you do all day?'

'Well, we've just bought another farm to move over to this system, so that's keeping me pretty busy. But while I'm working there, putting in the fences and all the rest, I know my land here is going forward, which is a really nice feeling. It takes a lot of work to get the system right, get the fences right, the grazing plans sorted. It takes a good deal of thought.'

He pauses.

'The thing is, in Australia and America we've absolutely pillaged our land. We've just fucked the whole thing. But I think we can turn it round really quickly.'

We're on the long drive back to Sydney and, ironically, it's raining. I turn to Tony. There's something I've wanted to ask ever since our conversation over dinner with Andrew and Megan at Etiwanda Station.

'You're going to buy farms, aren't you?'

'Oh yes.'

'How many?'

'We're starting with two and a half million acres. We're going to take large swathes of Australia and make it look right. Our minimum ambition is ten million acres. Then we'll move into other countries.'

'Can you raise the sort of money you'll need to buy the land?'

'We are working on just that.'

He's not kidding. I meet Tony in London some months later where he's holding meetings with large European pension funds who want to invest in long-term, sustainable assets. He's grinning like a Cheshire cat.

'Ah! Pomster!' he exclaims. 'Let me buy you a beer.'

Tony and Bruce want to save the planet. And they're going to make sure they have a whole heap of fun doing it.

CHAPTER 13

The President is busy

Men in exile feed on dreams of hope. AESCHYLUS

I'm looking at one of the strangest things I've ever seen. Eleven men are seated at three tables set in a horseshoe formation. The top table is reserved for a single occupant, clearly the head honcho (there are flags to either side of him). In the middle of the arrangement is an impressive piece of decoration: a huge piece of coral. Solemn business is at hand. A document is passed from person to person and signed by each. The signatures are slow and purposeful, indicating perhaps the seriousness of the text they are attaching their names to. Or are people just more deliberate about signing things when their role is that of a government minister? Or maybe it simply takes longer to write when you're dressed head-to-toe in rubber. And twenty feet underwater.

It's been an odd day. Two hours ago, along with a group of journalists and photographers, I'd stepped off a boat at Girifushi, one of the 1,190 islands nestled in the South Indian Ocean which make up the Republic of the Maldives. The sky, then as now, was a pure shimmering blue; the sand underfoot, soft and fine; the sea, a crystal clear aquamarine teeming with life. A cooling but light breeze had rustled through the palm trees which sheltered us from the sweltering sky. If it hadn't been for the armed soldiers observing us with a mix of slight disdain and bemusement it could have been a scene from paradise. On the quayside, executives and technicians bustled around a tent full of expensive-looking audio-visual equipment with wires running out the back into the lagoon. In the water, scuba-clad cameramen bobbed about waiting for instructions while intense looking *TV Maldives* staff rushed about with expressions that combined national pride with what I can only describe as the 'I hope we don't cock this up' look.

The eyes of the world will soon descend.

Girifushi, like all the Maldivian islands, is less than six feet above sea level. So, if seas rise quickly due to global warming, there could soon be a lot less of the Maldives to visit. That said, the islands have a natural advantage over continental coasts when it comes to resisting the waves, for like many island nations the Maldives are coral atolls. (The word 'atoll' is derived from *atholhu* in the local language, Dhivehi.)

Coral debris is constantly pushed onto the land by wind and water, and because the coral is always growing there's a continuous supply of island-building material. Paul Kench of the University of Auckland and Arthur Webb at the South Pacific Applied Geoscience Commission in Fiji looked at changes in the size and shape of twenty-seven atolls in the Pacific Ocean over a sixty-year period which saw sea levels rise, on average, about two millimetres a year. What they found surprised many. The land area of most had either remained stable or grown, and just fourteen per cent had lost landmass. 'It has been thought that as the sea level goes up, islands will sit there and drown,' says Kench. 'But they won't. The sea level will go up and the island will start responding.' That's the good news.

The bad news is that sea level rise appears to have been accelerating since the onset of the nineteenth century, and no one's sure how fast the coral can keep up. Then there are concerns over the West Antarctic Ice Sheet. If that becomes unstable (as some climate scientists fear it could), sea levels could quickly increase by ten feet. While we're at it, let's mention the melting of the Greenland ice sheet which has the potential to contribute another seven feet to sea levels if global warming continues unabated. There's no way for the atolls (or the rest of us) to keep up with that. Another worry for the Maldives is that a sharp rise in sea *temperature* could kill that island-building coral, leaving the nation without its natural defence against modest rises. Just such an event happened in 1998 when a spike in water temperatures devastated the nation's reefs (although there are now heartening signs of recovery). The fragility of the republic in the face of global warming is obvious.

This is one reason why President of the Maldives Mohamed Nasheed (the man sitting in full scuba gear at the top table) has become one of the world's most vocal and respected voices on climate change-related matters, and it is he I've come to meet. Nasheed argues that 'If we cannot save three hundred and fifty thousand Maldivians from rising seas today, we cannot save the millions in New York, London, or Mumbai tomorrow.' To the world he says 'we are all Maldivians now' and compares his country to Poland in the Second World War: 'a frontline state' in the battle against global warming.

But Nasheed isn't a pessimist, instead using the crisis to position the republic as a nation-sized laboratory of change, an example to the world of how we might arrest rising temperatures. 'The Maldives is determined to break old habits,' he told the UN. '[We are] no longer content to shout about the perils of climate change. Instead we believe our acute vulnerability provides us with the clarity of vision to understand how the problem may be solved.' Shortly after coming to power, he committed the nation to becoming carbon neutral within ten years. 'We do this not because we can solve global warming on our own. We do this because we hope to lead by example. If the Maldives can become carbon neutral, bigger countries might follow ... We are determined to formulate a survival-kit, a carbon-neutral manual that enables others to replicate it, in order that all of us together might just about save ourselves from climate catastrophe.'

Too much of the debate over climate change has been debilitating, 'about not doing things, about not emitting gas, about not going on holiday, about not having an ice cream,' Nasheed told *Al Jazeera* in April 2009. 'My feeling is this is the wrong way to go about it. We should be demanding we *do* things, do greener things, invest in renewable energy. Renewable energy is doable, it's feasible and will give you a handsome return.'

Others don't see the point. Earlier today Nasheed's communication adviser, the impossibly youthful Paul Roberts, told me, 'When the president announced the plan to go carbon neutral, we got a lot of pessimism in the international community, a lot of "Why bother? You're going to drown anyway."'

The president, however, is used to doing things against the odds.

Mohamed 'Anni' Nasheed has led an extraordinary life, and he's only forty-three. In 2008, his Maldivian Democratic Party, born during his exile in Sri Lanka and England, ended the thirty-year dictatorship of Maumoon Abdul Gayoom.

Nasheem had suffered under Gayoom. As he recalled, 'I was imprisoned on sixteen different occasions and spent a total of six years in jail. Of these I spent eighteen months in solitary confinement.' Stories of his confinement are horrific. His hands and feet were shackled. His food was laced with laxatives and broken glass. He was often placed in a cell deliberately designed to heat up like an oven in the baking sunshine. Gayoom's men tortured him 'to the brink of death' twice.

Gayoom's regime hired a London PR firm to present a tourist-friendly image. But Gayoom didn't do much to help. At a pro-democracy demonstration, he sent armed police to arrest a temporarily free Nasheed – who was subsequently charged with terrorism (the loudspeaker he was using to address the crowd was 'a weapon'). YouTube footage of the arrest soon made its way around the world (rather demonstrating Vint Cerf's point about the Internet's role in exposing violence) and the arrest sparked mass public dissent, building on tensions that had been growing over the previous five years. International observers voiced concerns that Nasheed would not receive a fair trial and, although it took another year in prison, he eventually walked free in exchange for a promise not to stir up revolution. But the revolution had already happened. Elections had been forced upon Gayoom and Nasheed became the nation's first democratically elected leader.

On taking office, the new president ordered the destruction of buildings previously used for detention and torture, and turned the multimillion-dollar presidential palace complete with gold-plated toilet (an opulent disgrace for a nation where one in five lives below the poverty line) over to the judiciary to house a new supreme court. And yet Nasheed allowed his former nemesis to walk free. 'A test of our democracy will be how we treat Maumoon,' he said.

Nasheed is a whole lot better at PR than his predecessor. As an ex-journalist, he has used his media savvy to bring the plight of the Maldives to the international stage. And his boldest PR stunt to date is the event I'm now looking at through my snorkel mask: the world's first underwater cabinet meeting.

I'm here, with a fair amount of the world's press, because the Maldives encapsulates the climate change issue. What happens in this island republic will be instructive to us all. It will be one of the first nations to cease to exist if sea levels rise too quickly. Indeed, Nasheed has already talked about setting up a 'sovereign fund' to buy land elsewhere should worse come to the worst. At the same time, the country is leading the way in creating a carbon neutral economy, which is no small feat when the nation is made up of over a thousand tiny islands, nearly all of which get their energy from diesel generators. Not to mention the fact that the Maldivian economy – and its carbon footprint – are dominated by tourism and associated aviation.

The Maldives also rather neatly demonstrates the carbon divide between 'developed' and 'developing' nations. Resorts renowned for their opulence host small numbers of wealthy tourists who consume luxury foods flown in to meet their demands. Sumptuous villas are filled with every electronic convenience. By contrast, neighbouring non-resort islands often have populations of thousands, many of whom live in poverty. 'We have a situation in the Maldives where you have a very poor third world island which is next to a very rich European island,' Nasheed has said – a legacy of the Gayoom era which saw too many tourist dollars going on the ruling elite in preference to social care. One of Nasheed's initiatives, therefore, is encouraging resorts to directly link their economies with neighbouring islands by buying labour and supplies.

Another reason for coming to see President Nasheed is that it allows me to take a step back from the science and consider, for want of a better phrase, 'the real world.' I've seen a good deal of techno-optimism, much of it well founded, but I want to talk to someone who understands more than most the emotional and political practicalities of change and how hard it can be. How, I wonder, can we turn knowing new things into doing the *right* thing? We may have solutions to our problems, but that doesn't mean we always use them. To paraphrase one of Winston Churchill's alleged sayings, 'Humanity can always be trusted to do the right thing ... once all other possibilities have been exhausted.' Perhaps a man who's gone from prisoner to president might have ideas on how to get us on the right path.

To save his own nation Nasheed needs the rest of the world to make changes to the ways they generate and consume energy. In an impassioned speech to the UN assembly, he was unequivocal: 'It's crystal clear to us ... If things go "business as usual" we will not live ... Our country will not exist.' In the same way he proposes linking the economies of resorts and neighbouring islands, he seeks to unite those who emit the most carbon with those who might be the first to suffer the consequences.

Due to some outrageous luck, I am one of just four people with no official capacity allowed into the (deliciously warm) water for the cabinet meeting. I swim around the perimeter careful to avoid numerous scuba-clad cameramen and wires trailing into the lagoon from the shore.

Underwater, cabinet ministers pass the document to each other in an orderly procession of aquatic cordiality, and one by one they use their waterproof markers to sign it: a call for nations around the world to cut greenhouse gas emissions that will be delivered to the forthcoming Copenhagen climate change conference. The odd familiarity of the exercise, I realise, is the thing that will make it great publicity. It is both ordinary and extraordinary simultaneously. Who *isn't* intrigued by a government meeting taking place underwater?

Because no one can talk, the whole affair is carefully choreographed using hand signals, with each participant following a waterproof 'order of service.' 'President signs statement. Cabinet pass the slate one by one. President signals cabinet ascent' ... and so on. Everyone sticks to the plan (I'm guessing a late amendment would be tricky), and given that most of them have had to take diving lessons in preparation I suspect they're just happy to get through the whole thing.

All in all the meeting lasts about twenty minutes. The ministers slowly rise from their watery seats and make their way back to the lagoon's edge. I find myself swimming next to the president. His head turns my way and I must look startled because he makes the underwater hand signal for 'Are you okay?' I signal back to assure him I am. I'm more than okay, but there isn't a hand signal for 'Bloody hell! I'm at an underwater cabinet meeting in the Maldives! *How cool is that?!*'

Reaching the shore, myriad camera lenses and microphones await us. The world's press are clamouring for the best vantage point at the water's edge, launching into a barrage of questions that Nasheed answers from the water, being careful to link the threat he sees to the Maldives with that faced by the rest of the world. There's also some good-natured banter about the benefit of having a cabinet meeting where none of your ministers can talk. Asked whether underwater meetings might become a regular feature of the Nasheed administration he replies, 'The whole idea is that this *doesn't* become a regular feature.'

It's odd to be in the middle of an international news event. I feel out of place bobbing around behind the president as the sun sparkles off the water, with possibly the biggest smile I've had on my face since getting a bike for my sixth birthday. On the short walk to lunch the president is waylaid by journalists eager to grab some time with him. I'm relaxed because my interview, scheduled for tomorrow, has been in the diary for months. Or so I thought.

Over lunch Paul Roberts, who deals with his international press and got me into the meeting, starts to use worryingly vague and expectation-limiting language. I'll 'probably' get my interview 'in the next two days', it's 'usually' not a problem, although the president 'has a very full diary'. My confidence is not bolstered when Paul suggests it might be 'helpful' to say a brief 'hello' to the president now 'just so he knows who you are.' Paul introduces me in a way that gives the strong impression this is the first time he's told Nasheed anything about me. I compliment the president on the day's success and say I am looking forward to our talk the following day. He looks confused. 'Are we having an interview?' he asks and Paul ushers me away.

In Paul's defence, today has been a huge media exercise and the man from New Cross is doubtless way down his agenda. Nonetheless, I've flown thousands of miles for the single purpose of interviewing a man with whom it seems I may not get to do more than trade one underwater hand signal each. A local journalist informs me that 'this kind of thing is common out here. You have to roll with it.' It seems like good counsel, not that I have much choice. For now, however, the sun is shining, lunch is good, people are smiling. There are worse places to be.

Next morning, Paul calls to tell me my interview won't happen today. 'I'm so sorry. None of us expected the reaction to the cabinet meeting to be so huge, the president is totally full up'. It's true that the underwater event has generated unprecedented levels of exposure for Nasheed. I've received emails from excited friends around the globe who've seen footage on news bulletins or read reports in their national papers. 'It's so great you're getting to interview the president!' writes one excitedly and I feel a knot in my stomach. The president is leaving for India the day after tomorrow. I take the day to walk the streets and the perimeter of the capital island of Malé, a view of the Maldives few visitors ever see (most transfer directly from the airport to their chosen resorts) and it's one that is revealing.

Malé's two square kilometres of land host somewhere between 90,000 and 150,000 people (definitive figures are hard to come by). Even by the lowest estimates this makes it one of the most densely populated cities on the planet. In stark contrast to the spacious whole-island resorts, Malé is a warren of tight noisy streets, filled with the buzzing of thousands of motor scooters. Buildings are crammed together, and it's hot, hot, hot. Looking at this mass of humanity crammed onto such a small island, my thoughts turn to the issue of population growth and its effect on the planet's resources. Here in the Maldives the population has increased six-fold in the last century.

In the year 1 CE, the population of the planet has been estimated at around 231 million. By 1804, there were one billion of us. But then we really got going. It took us just one hundred and twenty-three years to add another billion to the total. The next additional billion arrived in just thirty-three years. Fourteen years after (in 1960), we'd added another billion (to bring us to four). The next billions came quicker still, taking thirteen and twelve years respectively. By the turn of the millennium, we'd passed six billion. If we carry on like this, accepted wisdom is that the population of the planet will soon outstrip its resources.

By the docks I find an example of mankind's consumption. Hundreds of boats offload strong-smelling catches of huge yellow fin tuna. The fish are dumped unceremoniously into pots with their heads down (and thus obscured), their huge tails sticking into the air. In a tiled hall on the waterfront I witness the fish being gutted and prepared for sale. Huge knives expertly gouge out lidless eyes, heads are ripped off, spines are removed in single, swift and well-practiced movements by men smoking cigarettes and made indifferent to their butchery through repetition. I guess ripping the head off a tuna becomes as mundane as processing an invoice if you've done it enough times. Around me there is a huge amount of trade going on.

I ask one of the fishmongers if he was born in Malé and he says, 'No, but we move here because if you don't work on a resort, this is where the jobs are.'

Moving to the city to find work is the other key trend in population. Back when there was one billion of us, just three per cent lived in an urban environment. Today over half of us do. The UN's World Urbanisation Prospects report estimates that the number of us living in towns and cities will nearly double by 2050. 'Virtually all of the world's population growth will be absorbed by the urban areas of the less developed regions,' it says.

In 1798, foreseeing a catastrophe where human numbers would outstrip the carrying capacity of the planet, Reverend Thomas Robert Malthus famously wrote (and over nearly thirty years continually rewrote) his influential 'An Essay on the Principle of Population.' In it he predicted that food production would eventually fail to keep up with the population. 'It is an obvious truth, which has been taken notice of by many writers, that population must always be kept down to the level of the means of subsistence,' he concluded.

Malthus was both right and wrong. He was right because food production must, of course, keep pace with population if we are not to starve. He was wrong because so far, as the population has grown, we've managed to find ways to increase agricultural yields to match our needs. That is not to say that in some areas of the world there aren't shocking levels of malnutrition. This, however, is a problem with inequitable distribution, rather than production. Despite the dramatic increase in human population since the 1950s, world food production per capita has increased. That's a stunning achievement. There's more food per person than there's ever been. Or as the UN asserts:, 'There is a global food security situation that is steadily improving, with a consistently increasing global level of food consumption per capita.'

The big worry is that the effects of global warming will greatly reduce our capacity to grow food as crops wither under the onslaught of rising temperatures, which is why what Bruce and Tony are doing in Australia could be staggeringly important. Ironically, one way to continue to increase food production in a sustainable way could be to return to ancient practices. But can we continue to raise food production ad infinitum as our population continues to grow? The good news is we won't have to.

The statistician Hans Rosling from Sweden's Karolinska Institute tells a story of how many of his students wonder if keeping the poor alive is really such a good idea. 'In a time when we know the pressures on the environment are growing, my students tell me "population growth destroys the environment so poor children may as well die." This is a statement they don't make in class, but afterwards. They say, "Why save the lives of all these small children? Because if they survive, we'll be even more people, we'll destroy the environment [and then] we will all die in the end."'

The problem with this logic, says Rosling, 'is not that it's not moral, it's that it's wrong' – and here's why.

As the population is growing it seems to be stabilizing. The rate of growth is slowing down dramatically and looks set to stop (and possibly reverse) sometime around 2050. Already the population in many countries is falling. The important figure to keep in mind is 2.1. This is the upper limit of children, on average, every woman needs to have, to keep the population static. Between 1950 and 1955, the number of countries where the fertility rate was less than the crucial 2.1 figure was just five. In the period from 1995 to 2000, that number had risen nearly twelvefold to fifty-nine. According to the UN, countries who weren't replacing their population in 2010 included the United States, Canada, Russia, China, Australia, New Zealand, every country in Europe, Chile, Brazil, Sri Lanka, Thailand, North and South Korea, Tunisia, Japan and Singapore. The Maldives isn't replacing its population either, with a current fertility rate of 2.0.

Countries with the highest number of births and therefore rapidly growing populations are usually the poorest, with Africa dominating the table. Across Africa fertility rates of between 4 and 7 are common but crucially these too are dropping. Taken worldwide the human race's fertility rate has fallen from 4.92 in the 1950s to 2.56 at the beginning of this century. By 2050, the UN's 'median variant estimate' is that the figure will dip below the all-important 2.1 replacement level across the entire world population. This means we'll stabilise at just over nine billion people, and then possibly fall in number.

But why is this happening?

The contributing factors seem to be falling rates of child mortality, increased life expectancy, more urbanisation, improved access to birth control and the growing emancipation of women, all of which are in some way linked to increased prosperity. The first two reasons seem counterintuitive. If both us and our children are living longer, surely that would increase population, giving us more time to breed and creating more offspring to do the same? The figures, however, show the opposite trend.

One theory is that when you know fewer of your children are likely to die, and those that survive will live longer, you don't feel compelled to have as many. Second, prosperity is linked to increased choice, not only in whether to have children or not (through access to birth control) but also in what to do with your time. With the health of our children assured we can choose to spend more of our time being entertained and educated, and dedicating more time to educating and entertaining them too. With a few notable exceptions, prosperous nations seem to automatically reduce their birth rate (as the list of countries above amply demonstrates).

The rise of cities also has a role to play. One of the results of prosperity is urbanisation and this in turn also seems to encourage us to breed less. According to George Martine, lead author of the UN Population Fund's State of World Population 2007 report, eighty to ninety per cent of economic growth takes place in cities, which drives prosperity, which reduces the number of children we have. Martine also makes the point that the increased density of city populations 'makes it easier to provide social services, or services of any kind: education, health, sanitation, water, electrical power. Everything is so much easier, much cheaper on a per capita basis.' The report also found that female emancipation is more likely to occur in the city, again reducing birth rates. The city is where many go to make a career, often delaying starting a family. Property is also more expensive, meaning that even though you may be wealthier than many of the rural population, you have a lot less space in which to put your brood. And, of course, in the city the distractions of entertainment and education are ever greater. All this means that across the world, fertility rates are lower in cities than they are in rural areas – which, as the world becomes ever more urban, confers yet another downward pressure on the birth rate.

Beyond all this, cities are also good for biodiversity because they attract subsistence farmers away from the land. Subsistence farming might have a romantic image of honest rural folk in tune with their environment but the clue is in the name. Subsistence, by definition, is 'The action or fact of maintaining or supporting oneself at a minimum level.' Subsistence farming is characterised by extreme poverty, along with overgrazing and tilling of the land that has seen soil carbon levels plummet along with the reduction of biodiversity as people use the natural ecosystem for resources, a key example being the use of wood as a fuel for cooking and heating. When humans leave, natural ecosystems return.

The UN's State of the World's Forests report shows that as countries prosper and urbanise so their forests regrow. In Europe, forest resources 'are expected to continue to expand in view of declining land dependence, increasing income, concern for protection of the environment and well-developed policy and institutional frameworks.' In parts of Latin America 'where population densities are high, increasing urbanisation will cause a shift away from agriculture, forest clearance will decline and some cleared areas will revert to forest.' But the picture is less promising for low-income agriculture-dependent countries. In Africa, deforestation continues apace. Overall though, as we move to the city we leave behind ecosystems that regrow. A recent *New York Times* piece by Elisabeth Rosenthal, entitled

'New Jungles Prompt a Debate on Rain Forests,' went so far as to suggest that for every acre of rainforest being cut down, over fifty acres of secondary forest are regrowing. That's pretty incredible, don't you think? It is, however, important to stress that re-grown secondary forest in no way matches up to the biodiversity lost when ancient rainforests are destroyed.

This teeming metropolis I'm walking through is therefore an engine of renewal. Hans Rosling's students are dead wrong. Those 'poor children' must live, move to the city and prosper – and in doing so they will, in just a few generations, stabilise the world populace while allowing many ecosystems to flourish.

It's a nice thought to have as the sun sets on the Indian Ocean.

The next morning brings fresh disappointment. An interview today is looking 'unlikely,' says Paul. He suggests, instead, that I come along to a lecture the president is giving about Gandhi, and try and talk to him there.

At the lecture I sit next to a man called Per, who turns out to be a) recovering from Dengue fever and b) the head of the Red Cross and Red Crescent societies in the Maldives. On my other side an aggressively cheery fellow called Wahid makes easy conversation, laughing and smiling with each inhalation and exhalation of breath.

As the lecture ends I seek out Paul who directs me towards Nasheed's right-hand man, the former democracy campaigner Mohamed Ziyad ('You'll recognise him from his ponytail').

I find my man at the buffet lunch and introduce myself. 'I'm hoping to interview the president,' I say and recount my months of communication.

He assesses me with a kind of bemused indifference. 'This is the first I've heard of you,' he says. 'There is no chance of you getting an interview. The president is busy.'

And then I spy the jocular Wahid, nibbling on spring rolls and ... talking to the president. They're an odd pairing. Nasheed's slight and diminutive frame gives him the countenance of a jockey, while Wahid looks like a Maldivian Oliver Hardy. I use the fact that I 'know' the larger man to infiltrate the circle, via a Ziyad-distracting 'dummy' visit to the buffet (where I do admittedly pick up an onion bhaji).

I compliment Nasheed on the lecture and remind him of our brief introduction after the underwater cabinet meeting. Ziyad is moving towards us. I make my last ditch effort, explaining (very quickly) to Mohamed Nasheed that I have travelled here on the promise of an audience with him and will leave his nation before he returns from India.

Ziyad is now with us. This is absolutely my last hope. The president turns to him. 'My diary is pretty full today?' Ziyad nods. 'So, the only way we can do it ... is now? We can do it now I think, but let's move to my offices.'

Ziyad looks annoyed. But before I know it I'm in a coterie of officials (including, I notice, security staff with those funny earpieces) being escorted out of the building. Ziyad ushers me into the back of a black-windowed car that starts to drive off before I'm fully in it and he reprimands the driver. From this moment on he becomes helpful, if still rather peeved. We arrive at the presidential offices and rush straight through security.

In the lift I try to break the emotional stand-off between us by asking him if he was with the president during his exile in Sri Lanka (where Nasheed suffered an assassination attempt) and Britain (where he was granted asylum).

'No, I was here,' he says.

'That must have been difficult?' I respond. 'The last regime didn't really make it easy for you.'

He looks at me like I've just said the most facile thing possible. And then his face both softens and saddens a little. 'It was hard,' he says quietly.

That's an understatement. I subsequently find out that the businesslike man escorting me to my interview was so severely treated by the National Security Services that it took a long spell in intensive care to recover. Now he confidently walks the corridors where the former regime endorsed and ordered the indignities he was forced to suffer. I'll probably never get a chance to talk to Mohamed Ziyad again and that's a shame. His story, like so many who fought for change here, is extraordinary.

But I'm ushered into a wood-panelled meeting room and a few minutes later Nasheed enters, smiling. It's less than fifteen minutes since Ziyad had told me there was no chance of me getting an interview.

Close up, the first thing you notice about Nasheed are his eyes. They seem infused with light and I get the feeling that behind them and at every moment a great deal of work is going on, as if they're small windows that overlook a huge intellectual conveyor belt. He wears a smart but not flashy suit and, despite the heat, a tie. The impression is one of care rather than fastidiousness. But perhaps most surprising is his candour: there is little of the guarded talk typical of career politicians. Early in our conversation he says, 'You know I'm always told "be cautious – not to do that, not to say this." I end up saying something "wrong" every week and they don't like it.' He smiles. 'But I have to go on saying what I believe in.'

The 'they' in question is the governmental machine he's inherited, which clearly frustrates him. How does he cope? I ask. He laughs and says something I really don't expect.

'I think what's helping me is Tom Sharpe' – an out-of-the-blue reference to the English comic novelist, famous for graphically and lewdly lampooning authoritarianism. 'The comedy of it all – of government and endless meetings. You'd be amazed at the kind of "work" I do,' he chuckles.

The Copenhagen climate change conference (which Nasheed attends two months later) must have been a special kind of hell, yet he was hailed as 'the real hero' of the conference by Danish Prime Minister Lars Rasmussen. 'The Copenhagen Accord is a long way from perfect. But it is a step in the right direction toward curbing climate change,' said Nasheed, before returning home to get on with leading by example.

I get the impression that here is a man who understands better than most how to negotiate the seemingly impossible, and who has the patience to endure and guide an almost infinite progression of baby steps to a resolution. As he says: 'When you reach a dead end in trying to convince someone or trying to do something, it would be best to give it a moment, and give it some thought. There's no value in just banging on. There's always more than one avenue to any destination. Even in a dead end, when things get really, really bad you have to keep going – however badly you suffer, whatever losses you incur, you just have to keep going. You have to make a tiny step. Don't just be stuck with a single option.'

'Are there parallels between a man in solitary confinement and a tiny nation in the midst of the world's biggest problem?' I ask. 'Do you think something in your solitary confinement prepared you for this role?'

He looks at me squarely and I worry for a second that my question might be taken as making light of his ordeal, that I'm trying to spin it as a useful experience and have therefore trivialised what happened to him. But instead he exclaims, 'I think you're very right there! Yes! If you can muster the faculties to survive in solitary for long periods of time, you must have some mechanisms, some tools upon which you can build a strategy for stopping global warming. Very true.'

He points, not at me, but as if at some philosophical target hovering between us. 'I was just one person right in the middle of some huge, very sophisticated machinery. And we as the Maldives are in solitary, surrounded by bigger nations and big countries with huge achievements and we are just probably nothing, but still . . .' He shrugs. 'We have our ideas. We want to survive. We are not asking for much.'

From this point on it becomes hard for me to separate the man from the nation. When he speaks, 'I' and 'we' become interchangeable. When he recalls the struggle for democracy or talks of the current battle against global warming, his language is also transposed freely between the two (indeed, he often slips between time frames when answering a question). For him, it seems all events are linked in a continuous mission to help his homeland flourish. Perhaps this is why so many people find Nasheed a compelling negotiator – an ability to come across not just as a representative of his country, but an embodiment of it. In our time together I certainly begin to get this feeling. It's as much in the way he rebuffs hard-line Islamists (who have publicly criticised him this week for removing his wetsuit and exposing his chest at the close of the underwater cabinet meeting) as it is his views of political and environmental realities. 'No sane Maldivian would think you could be in the water with anything on you. What are they talking about? Have we ever gone swimming with a T-shirt on? No! So why should the president? That is not the Maldives.'

Climate change is, he says, 'the biggest challenge we will ever face. Not terrorism, not piracy, not drug dealing; nothing compared to this. So we really need to try and do something about it. No matter how small or however insignificant we may be.'

Given the severity of the threat he sees to his nation, how does he remain optimistic?

'With the belief that there is hope, that there is a bright future. By seeing a picture other than the very fearful picture that is staring you right in the face.' He slips time frames and we're back in the democracy battle. 'What I would try to do is imagine another country, another homeland, another time, other circumstances.'

'It's all about having the right vision?'

'I've always been optimistic. I feel if you can show the light at the end of the tunnel it's bearable to move out from difficult times and situations. I know this is huge odds, but if you look at the situation the Maldives were in five years ago, most people would have said, "What is the point with the democracy movement? We are wasting our time."'

The fact that I'm having this conversation with a head of state who was a former political prisoner, while sitting in the presidential office, amplifies the essence of his argument.

The big picture, the goal, the brighter future, optimism that things can be achieved. All admirable ideals but how, I ask him, does one hold on to these when you're embroiled in the 'endless meetings' and the 'comedy' of government he talked about earlier?

By way of an answer, he sighs and puts his hand to his temple in a gesture of comic resignation. This inspires me to ask if, in a strange way, he misses his days spent under house arrest where, if nothing else, he had plenty of time to think. He smiles.

'I really do,' he says almost wistfully. 'I'm surviving from the reserve – and my feeling is one might only be able to survive for five years on that – and this is probably why five years is a natural term for anyone to be a leader.'

'Your brain is too full with the day-to-day now?'

He leans forward. 'You don't get the bigger picture,' he says urgently. 'You lose the concepts. So then you get hold of processes, frameworks, strategic plans, matrices – it's all very good . . .' He trails off. 'But if you don't see the bigger picture . . .'

This is the essence of leadership. Keeping to a simple vision, even when things get complicated. I'm reminded of something Bruce Ward said to me in the Australian outback: 'Keeping things simple is never easy.'

The president points to the clock indicating that our time is nearly up. I decide in the minute or two I have left to do a bit of advocacy for Klaus Lackner and his carbon scrubbing technology. As I explain the potential of Klaus's work I can see Nasheed's brain working. His eyes go up and to the right. He leans forward. How much money does Klaus need? (Nasheed tells me he is hoping to put aside a hundred million dollars each year for investment in carbon neutral projects). I repeat the figure Klaus gave me of twenty million dollars to take his technology to the next stage – a design that can be rolled out worldwide.

Suddenly I find myself suggesting to the president that maybe one of the deserted islands in the Maldives might act as a good demonstrator for the technology, to show the world its potential. As an act of cheek it goes beyond anything I've ever done (clearly this is a day for sticking my neck out). Except the president says, 'So he should come down here, and we could give him some room . . .'

Should I put him in touch with Klaus?

'Please do!'

And what would be the best way of doing that, I wonder?

'I think the best way would be to get in touch by personal email,' he says, scribbling his contact details on a piece of paper and handing it to me. I'm dumbfounded. A president has just handed me his private email address (hosted by one of the world's popular web-mail providers).

I have one last question. What's his one tip for approaching the future?

'Never give up hope, you know? Never give up. Just keep moving.' He pauses. 'Tomorrow must be better. Tomorrow is better.'

Back on the humid streets of the capital, I have a huge grin on my face. It's not just that I finally got my interview, it's because the interview has made me feel lighter. Nasheed is a lightning rod for optimism, and it's hard not to feel better about the future after spending time with him. And then I realise that it's not just him, but all the people I've been meeting on my journey. All of them are inspired by what can be done, all of them are doing something to make it real, all are motivated by a desire to improve the human condition and none of them is waiting for permission.

During our talk the president had said, 'Thoughts are real. Once you have given thought to something, it then becomes material very often and quickly.'

He is, of course, right. A few weeks after I leave the Maldives, Nasheed announces the government has commissioned an offshore wind farm comprising thirty large turbines delivering power via a network of underwater cables. When it's finished, the plant will provide forty per cent of the nation's electricity and reduce its carbon emissions by a quarter.

You get the feeling he's only just getting started.

I'm not used to luxury. Or rather, I'm not used to the luxury of the rich. (I understand the luxury of a really good Yorkshire pudding as much as the next man.) And yet here I am at the Soneva Fushi resort on Kunfunadhoo Island. Regularly voted one of the best resorts in the world, it's the *crème de la crème* of holidays for the rich and famous who want to get away from it all and enjoy a quiet, understated opulence. I could never afford to stay here, but am getting a glimpse of how the other half live as a guest of the owners Sonu and Eva Shivdasani. Arriving on the island, I'm assigned my own personal 'Man Friday,' the impossibly smiley Hannan who tells me that the maximum occupancy of the resort is about a hundred and fifty guests who sleep in private villas (each with their own swimming pool), attended to by a staff of nearly four hundred. The smaller neighbouring island of Maahlos, known locally as 'Mosquito Island' and where much of his family live, has a population of three thousand.

On one level, Soneva Fushi could be seen as everything that's wrong with the planet – rich westerners with massive individual carbon footprints holidaying next to unseen poverty. But the hotel chain that owns the resort has committed itself to be carbon *negative* by 2020, and that includes offsetting the flights of all its guests via a scheme to finance community wind turbines in southern India.

The tall and willowy Anke Hofmeister, Soneva Fushi's resident marine biologist and environmental manager, takes me on a backstage tour.

'This is the best job in the world isn't it?' I ask Anke.

'Yeah. Except there are no single men. Other than that it really is,' she replies, smiling.

Anke shows me how waste packaging and food is mixed with the island's sandy soils, increasing their carbon content and allowing the island to grow most of the vegetables used in the kitchens and cutting down on air miles while providing super-fresh produce. The remaining food waste decomposes in a 'biogas plant' producing fuel for cooking and power generation. Non-food waste is turned into biochar which is used to improve soil fertility further. Waste glass is crushed to make building material. Anke shows me how excess heat from diesel generators is used to warm water for guests. A small solar farm being installed during my stay will soon provide ten per cent of the island's energy needs, cutting down on that diesel.

It's all pretty impressive, almost a microcosm of the solutions to global warming I've been investigating. 'Of course,' I reason, 'it's easier to be a responsible corporate citizen when your clients are the super-rich.' There again, if every resort in the nation followed this example, the republic would already be close to achieving Nasheed's objective and I'm suddenly struck by the ugliness of my own cynicism.

I wander off to look at the sea and have a think. While I was in Boston, I'd caught a one-woman show by Gioia De Cari about her time as a maths student at MIT: 'Getting an education from MIT is like taking a drink from a fire hose,' she'd said. That's how I'm feeling about my trip. I expected the future to be different, but I didn't expect it to be so radically different. I'm disoriented in paradise. I've seen revolutions in medicine, biology, robotics, nanotechnology, energy production and communications that could change everything, if we can avoid a climate catastrophe (and I've glimpsed solutions to that, too).

I'm overwhelmed with new knowledge. How does all this fit together? Where are we going? How do I find 'the big picture' and then hold on to it – as even a man like Nasheed finds a battle? How can we make the right choices? I started this journey because I wanted to find the future but suddenly I'm feeling lost inside it. To help make sense of it all, I'm going to meet four people who live right in the middle of the future storm of possibilities I've seen coming down the line.

And then I want to go home.

PART 4

RE-BOOT

CHAPTER 14

Making a road where there isn't one

Are we nearly there yet?

During my visit to Washington, DC, to meet Google's Vint Cerf, I used my one evening in the city to tour its monuments and historical buildings. I stood outside the White House and asked a fellow tourist to take my photo, I watched children try to touch as high as they could on the great needle of the Washington Monument, and I walked the steps of the Lincoln Memorial.

On opposite walls of the memorial are two of President Lincoln's most famous speeches: the Gettysburg Address and the Second Inaugural Address. But my favourite words of his do not appear here. They were written in his annual message to Congress in 1862, when the country was in the midst of civil war. These words have come back to me now with a new resonance:

> The dogmas of the quiet past are inadequate to the stormy present. The occasion is piled high with difficulty, and we must rise with the occasion. As our case is new, so we must think anew, and act anew. We must disenthrall ourselves, and then we shall save our country.

Another fan is educator Ken Robinson who quoted these words at the 2010 TED conference. 'I love that word, "disenthrall,"' he said. 'You know what it means? That there are ideas that all of us are in thrall to, which we simply take for granted as the natural order of things; the way things are.'

Sitting on the beach on Kunfunadhoo Island, I was struck that the way I think and reason is in thrall to a world that is passing. Who knew that we might be able to cheat death, that we can harness the very code of life itself, that machines might be thinking and feeling, that we can manipulate matter with molecular precision, that the Internet has barely got out of the starting blocks, that data is becoming as real as flesh, that flesh and machines can merge, that we can make fuels from sunlight and air, that we can manage the climate, go off-grid, and harness the sky to remake the Earth's soils?

I mean, I had a vague idea that *things were happening in laboratories*, but until I set out on this journey I hadn't comprehended that it was *really going on*. But now I feel it. I've seen and touched something of the world coming and met with its creators. The opportunities open to us to improve the world are buzzing around my head and at the same time I worry about my – and our collective – ability to grasp them.

My personal problem is that there was nothing in my 'learn-enough-facts-to-pass-the-exam' education that taught me how to deal with what I've found out. Sure, I can absorb information, even make sense of how some of it fits together, but I've a strong feeling that I'm missing something. The subjects I've investigated on my travels are not only a subset of current innovation, they are a tiny fraction of innovations we cannot yet imagine. Nobody was talking about mobile phones or data networks at the 1878 Paris Exhibition where Augustin Mouchot showcased his solar-powered ice-maker.

I need to make a final step and let go of ways of thinking that are so ingrained in me that I may not even be able to see what it is I need to change. So I've come to see some people who I hope can disenthrall me – individuals known for their ability to think outside the box while living in it; people who have spotted something of the big picture. I don't just want their best guess of what the future may look like. No, what I'm looking for now are *ways of thinking* about the emerging world that will enable me to handle and comprehend something of the march of history to come.

During the Enlightenment of the eighteenth century, reason and faith battled it out and, in many countries, came to a workable compromise – a mix of secular governments, which separated religion from the business of the state, and liberal democracies that asserted the right of individuals to follow their faith unhindered. At its heart was a belief in science and rationalism, and a critical questioning and re-evaluation of the world. I'm wondering what questions we need to ask now of our institutions, customs and morals. It's not that we don't have the physical tools to remake our world for the better (I've seen them), but what are the mental ones?

I need a re-boot. And I'm about to get one.

My first port of call is a return trip to Boston. I'm met at the airport by Tracy Wemett, Konarka's PR liaison. Tracy has generously offered me the use of her basement room for my stay, which makes a welcome change from hotels. Though first we've got to get to her apartment alive which, given her driving, is not a given.

Since I saw Tracy last, it seems I haven't been the only one to notice her maverick approach to the road. One speeding ticket too many and she's been required to take a driving education course by the State of Massachusetts. The results are reassuring. 'I was told I'm the sort of person who likes making a road where there isn't one.' She pauses. 'Apparently that's not good.'

I'm back in Boston chiefly to see Ray Kurzweil, variously an inventor, guru, madman, prophet or genius, depending on who you listen to. For some he's the world's leading futurist – the man who really gets it. Others have accused him of being the ghoulish high priest of a techno-cult that has more in common with science fiction than serious analysis.

I first came across Ray's ideas back in Oxford at the start of my trip. Ray, like Nick Bostrom of Oxford University's Future of Humanity Institute, is a transhumanist who intends to live for hundreds if not thousands of years. I bumped up against his ideas again whilst investigating robotics and artificial intelligence at MIT and Cornell. Ray believes the line between humans and machines will blur and we will soon leave behind the shackles of our cruel biology and evolve, through a merger with technology, into a new species, or set of species – posthumans with powers that we cannot yet comprehend. He popped up in my conversations with Eric Drexler (mastering nanotechnology is fundamental in shaping the future Ray foresees) and Vint Cerf at Google (Ray was one of the few futurists to accurately predict the rise of the Internet).

Throughout my journey I've been reading his book *The Singularity Is Near* – which talks of a soon-coming epoch 'The Singularity,' 'a period during which the pace of technological change will be so rapid, its impact so deep, that human life will be irreversibly transformed.' Depending on who you speak to, the Singularity is either pure bunkum, the next great leap forward, or a dystopian apocalypse where posthumans will enact a genocide on their less developed forebears. Ray thinks it'll happen sometime around the middle of this century.

Kurzweil Technologies occupies one floor of a nondescript office block in the town of Wellesley, about fifteen miles from the centre of Boston. I take a seat in the cramped reception area while I wait for the man himself. It's a slightly surreal place with a full size waxwork figure sitting opposite me, dressed in a suit and wearing a badge that announces 'I am an inventor.' (I later find out he's called George.) Standing forlornly in a corner behind a table displaying Ray's many books is one of Thomas Edison's early dictation machines. The walls and tables are festooned with awards too numerous to list (I estimate at least thirty) but which include the US National Medal of Technology. Ray is a busy man. It's taken the entire duration of my journey to arrange a time to meet and even now he keeps me waiting for over an hour. Eventually we take a seat in his office.

He has an air of extreme calm. When I mention his critics, his defence is robust but his voice doesn't alter. He deals with every point in a cool, measured way. He seems to have his emotions perfectly house-trained. Watching speeches by Kurzweil is often compelling, but never because of his delivery. He'll say something that'll make your head spin in the same tone he might use to order a sandwich. I tell him about my trip and the people I've seen. He seems satisfied with the roll call. He's collaborated with Nick Bostrom, nanotechnologist Eric Drexler is a friend, and Cynthia Breazeal appears in Ray's docu-drama film version of *The Singularity Is Near.*

Ray's journey to visionary genius/techno-prophet/crazy person (delete as appropriate) stems from his work as an inventor, in which he has a fine pedigree, as evidenced by all those awards in reception. Among other things, he invented the first machine that could scan text in any font and convert it into a computer document – a technology he applied to building a reading machine for the blind. Stevie Wonder was the first customer and became a friend, which in turn led to Ray inventing a new breed of electronic synthesisers that were able to capture the nuances of real instruments. In a former life as a musician, I coveted the Kurzweil K2000 but was never successful enough to afford one.

'Most inventions fail not because we can't get them to work,' he says, 'but because the timing is wrong. So I started looking at information technology trends.'

What Ray found was something extraordinary – a clear, unmistakable pattern of information technologies doubling (or more) their performance while halving (or more) their cost at regular and predictable intervals. Those intervals are different depending on the specific technology – some might double in six months, some might take two years but that's not the important thing to keep in mind. What's important is that they'll take the same amount of time to double again. Most of us are familiar with this trend in the way computer-processing power has skyrocketed ('a billionfold since I was a student,' he says) but this rapid doubling can be found operating in many other places too.

Its significance stems from the fact that each tool we build gives us a better platform on which to build its successor. Computers, for instance, allow us to design more powerful computers than themselves. This phenomenon is called 'autocatalysis,' where the output of a process can be fed back into the process itself, spurring it on. In the opening chapter of *Whole Earth Discipline*, Stewart Brand gives a useful perspective:

> Not all technologies are autocatalytic: New discoveries don't make every technology advance faster. Progress in automobile technology and wind technology makes better cars and wind generators but not better tools for the engineering itself. The current autocatalytic technologies that goose themselves into exponential growth are infotech (including computers, communications, and artificial intelligence), biotech, and nanotech (which is blurring into biotech). What's more, they stimulate each other in a mutual catalysis that at times results in hyper-exponential growth of power.

Ray calls this phenomenon 'the law of accelerating returns' and what it means, he says, is change will come faster than we think.

Author Matt Ridley is fascinated by what he calls 'the rapid, continuous and incessant change that human society experiences in a way no other animal does.' He puts this down to the fact that human culture allows ideas to meet and interact, something he calls 'mating minds' or 'when ideas have sex.' It's another perspective on Robert Wright's 'nonzero-sum game,' the idea that trade and the sharing of culture across boundaries joins our fates together as we come to depend on each other's specialities and ingenuity.

From everything I've seen, ideas like to have sex a lot. They're not so much squeezing in the occasional fumble as going at it like bunnies. One of the things that has been getting them in the mood, according to Ray, is the power of information technology. As it increases in power, it also infiltrates all disciplines, helping us process more data, find fresh knowledge and build new tools – tools that make the next round of innovation quicker and cheaper. It helps us find ways to process genome sequences faster than we could just a few months ago, design new materials with more precision than last year's efforts, see deeper into the atom and further out into space, connect billions of us together and put processing power into every corner of the biosphere.

To understand the implications of the law of accelerating returns, you have to get your head around how potent the effect of doubling can be. Think of it this way. Let's say you travel a foot with each step you take. If you take ten steps, you'll have covered ten feet. Now imagine that instead of each step progressing one foot, you somehow double the distance you covered with the last one. So while your first step covers one foot, your second covers two and by your third step, your stride is four feet. The difference between normal-stepping and doubling-stepping is extreme and gets ever more so. As a doubling-stepper, you will cover not ten feet in your first ten steps, but over a thousand.

By the time you've done just twenty-seven steps you'll have travelled more than the distance around the equator. At this rate you could walk to the sun and back in forty steps (your last step having covered 549,755,813,888 feet). One can only imagine the trousers you'd need. Meanwhile, normal-stepping has got you about one-tenth the length of a football pitch. Now, of course you can't step like that, but technology can.

Thinking back, I've seen numerous examples of mankind's exponential adventure on my trip: from the plummeting cost of genome sequencing to the ever improving 'cost per watt' performance of solar technologies. Ray cites these examples and others. The first hundred pages of *The Singularity Is Near* almost bludgeons the reader with graph after graph, based on historical data showing exponential growth in the number of phone calls per day, cell phone subscriptions, wireless network price-performance, computers connected to the Internet, Internet bandwidth and so on. These all have a computing flavour, but Ray sees exponential growth of knowledge too, using rocketing numbers of nanotechnology patents as an example. Ray quotes example after example because he wants us to get past what he sees as an inherent prejudice in our thinking.

'Our intuition is linear and I believe that's hardwired into our brains. I have debates with sophisticated scientists all the time, including Nobel prize winners that take a linear projection and say "it's going to be centuries before we . . ." and "we know so little about . . ." and here you can fill in the blank depending on their field of research. They *just love* to say that. But they're completely oblivious to the exponential growth of information technology and how it's invading one field after another, health and medicine being just the latest.' He continues, 'The evidence for exponential growth of information technology is very persuasive. It's the most startling thing, how smooth and predictable it is. It doesn't progress in fits and starts, which is what you'd think and what I expected. The graphs continue on a very smooth trajectory.'

Ray says that what those graphs are actually measuring is innovation, creativity and competition.

'You would think that those are the least predictable aspects of human behaviour and indeed each individual project is unpredictable, but the overall results are foreseeable to a high level of precision. That's because somewhere someone always makes the leap that keeps the exponential going.'

I find myself nodding. I've only witnessed a tiny slice of the innovation going on around the planet but I've seen enough leaps forward to get a sense of what Ray is saying. The strongest example for me is that of Cornell's Hod Lipson, who told me about the machine he'd made that could find truths from data. 'We can go from data *straight to laws*,' Hod had told me. 'Previously people could only go from data to predictions. So now a scientist can throw in some data, go and have a cup of coffee, come back and see fifteen different models that might explain what is going on. That saves a lot of time. Previously coming up with a predictive model could take a whole career. Now at least you can automate that so you can focus on meaning.'

A machine that shortens the discovery of new knowledge by the length of *a career* is the very thing Ray is talking about: an advance in knowledge that speeds up advances in knowledge. But since I saw Hod, the machine has gone one step further.

When I'd left Cornell, Hod and his team were battling to understand an equation derived from the observation of the soil bacterium *Bacillus subtilis*. The mechanical brain had come up with an equation no one could make sense of. Sure, it had taken less time to come up with a law, and this was indeed allowing the team to concentrate on meaning. The only problem was, they weren't finding any. Hod wondered if it was hopeless, if the machine had given him the equivalent experience to that perceived by a dachshund being read Hamlet.

Now, as I near the end of my travels, I've received an excited follow-up email: 'We finally understand what the equation means. It's fantastic!'

How had they found out? Long hours scratching their heads over cups of coffee and biology textbooks? No, they worked out a way to get the machine to explain it to them. Hod and his colleagues, Michael Schmidt, Gurol Suel and Tolga Çağatay, asked the brain to conduct a new set of experiments on the results of the *Bacillus subtilis* investigation. These new experiments set about finding relationships between the mystery results and existing knowledge, shedding new light on the meaning of the equation. In short: robot finds out new knowledge and explains it to its own inventors.

The pace of finding stuff out just racked up another notch. And not just at Hod's laboratory. Our ability to churn through and analyse massive amounts of data is having dramatic impacts on everything, including, notably, medical research. Rather than coming up with a hypothesis for what causes a condition and then spending years in painstaking medical trials to see if they are right, some researchers are now processing vast amounts of existing medical data and looking for patterns that might pinpoint contributing factors to disease. It's an upside-down way to do science that we're not yet used to. Data first, hypothesis second.

George Church's Personal Genome Project will provide one such dataset. Another is emerging from Google founder Sergey Brin, who is funding a pattern-finding project to assist in the cure for Parkinson's disease (which analysis of his DNA tells him he has a 30–75 per cent chance of developing). 'Generally the pace of medical research is glacial compared to what I'm used to in the Internet,' Brin says. 'We could be looking lots of places and collecting lots of information. And if we see a pattern, that could lead somewhere.' So he recruited a group of 10,000 Parkinson's sufferers, had the company 23andMe (which is largely funded by Google) run their DNA, and set out to find links. It's one of the many examples Kurzweil cites of information technology 'invading one field after another'.

Sitting in front of Ray Kurzweil, I'm getting just what I came for. I'm becoming disenthralled from my inclination to think linear. We must understand the power of the exponential, he urges. If we don't, progress will outrun us, and our personal decisions will be hopelessly out of step with an unfolding reality. The good news, Ray suggests, is that this ever faster rate of progress is good for the planet.

'All these concerns that we're running out of resources would be absolutely true if the law of accelerating returns didn't exist,' he says. 'For instance, people take current trends in the use of energy and just assume nothing's going to change, ignoring the fact that we have ten thousand times more energy that falls on the Earth from the sun every day than we are using. So if we restrict ourselves to nineteenth-century technologies, these concerns would be correct.'

In other words, the law of accelerating returns should soon see a green energy revolution as solar power keeps doubling its efficiency and halving its cost, leaving fossil fuels standing. After my experiences seeing organic solar cells rolling off a printer just a few miles from here, I'm inclined toward his point of view. I've seen the kind of innovation that keeps the exponential going.

In 1999, Kurzweil published *The Age of Spiritual Machines* in which he applied his understanding of the law of accelerating returns to make predictions, and – handily – made a bunch for 2009. Critics and advocates alike have leapt on these, loudly proclaiming 'Ray was right!' or 'Ray was wrong!' often in the service of their own hopes and fears rather than reasoned analysis.

According to Ray himself, he made 108 predictions of which 89 are correct, 13 are 'essentially correct,' three are partially correct and two are ten years off. Just one is wrong, he claims but that was tongue in cheek anyway, as it predicted that the two most popular choices for intelligent computer assistance would take the form of 'Maggie, who claims to be a waitress in a Harvard Square café, and Michelle, a stripper from New Orleans.' I counted his predictions slightly differently and concluded he got nearly two-thirds right, including forecasts about how we'd interact with media, the power of computing, the rise of networks, the growth of wireless connections, the role of technology in tackling disabilities, advances in nanotechnology, the issue of privacy as a political hot potato and improvements in medicine.

Of the rest, I put about half in the 'sort of right' category, by which I mean that the prediction has come true, but not quite in the form Ray describes it, or it's something that looks like it's coming soon (for instance, 'the greatest threat to national security comes from bio-engineered weapons'). The remainder are 'wrong' but only in that Ray was optimistic on the time frame. For instance, critics tend to leap on his prediction that 'translating telephone technology (where you speak English and your Japanese friend hears Japanese, and vice versa) is commonly used,' which hasn't happened, but one must bear in mind that exponential growth may look small at the outset. Doubling from 0.1 per cent to 0.2 per cent is a small step, but when things start leaping from a thousand per cent to two thousand, they seem to 'come out of nowhere' even though the exponential trend has stayed constant. Translating telephone technology could arrive with just such a bang at some point in the near future.

It seems, in fact, that I've spent my journey seeing much of what Ray forecast in 1999 coming true.

If Ray's right about the law of accelerating returns, and I think when it comes to infotech, nanotech and biotech he is, this means that when I decided to investigate the future of my lifetime I was being hopelessly over-optimistic. What I've actually been exploring is the next ten or twenty years at best. Everything I've talked about on my journey – from Eric Drexler's nanofactory, to the biological immortality of Nick Bostrom, to George Church's 'personal genomics,' to Cynthia Breazeal's sociable robots – is coming quicker than we expect, along with synthetic fuels and a solar power revolution. By the time I'm one hundred, the ever increasing rate of change will have made whatever is coming literally indescribable today. Trying to write a book about that would truly be an exercise in science fiction.

It's talk of such rapid technological evolution that brings Ray into conflict with other scientists and commentators who think he's not so much running before he can walk as getting into the cockpit of a jet fighter straight from the crib. For Ray, though, that's kind of the point – crib to jet fighter is really just a few doublings, the law of accelerating returns in action. In fact, Ray points out that sometimes the rate of doubling can double itself, creating the 'hyper-exponential growth' Stewart Brand references. Others are unconvinced. They see Ray the same way the State of Massachusetts sees Tracy – he makes a road where there isn't one. Douglas Hofstadter is one critic. Now a cognitive scientist at Indiana University, he most famously authored *Gödel, Escher, Bach* – an attempt to explain how consciousness can arise from a system, even though the system's component parts aren't individually conscious. Hofstadter told *American Scientist* that he thought Ray's ideas were like a blend of 'very good food and some dog excrement' that made it hard to untangle the 'rubbish' ideas from the good ones.

That's gotta hurt.

My sense is that Ray plays with ideas that touch on some of our deepest hopes and fears, and that's why people react.

'Take death,' says Ray, which is a hell of a way to open a sentence. 'We've spent thousands of years rationalizing death. There was no argument you could make for defeating it until fairly recently. People still disagree but you can make a very credible argument that we can defeat death, not in the mystical way that traditional religions give us by saying death is ennobling and we live forever on the other side, but by using technology. All of our institutions are going to have to be rethought. Marriage, for instance, is based on a premise that people aren't going to live that long.'

There you go: death, religion and sex in one brief argument. It takes enough effort for our linear brains to get themselves around just one or two of these ideas and then Ray comes along and says, 'Belt up, things are going way faster than you thought, and by the way that means I don't intend to die. Would you like to transcend your biology with me and embrace the next stage of techno-human evolution? Hurry now.'

It's no wonder some people find Ray just *too difficult* to engage with. In my research, I've found myself reacting strongly to his ideas too. But what I am taking away from him is the understanding that coming to grips with the exponential growth of technology is a key part of my personal re-boot.

I want to ask Ray a wider question, though. 'Have you got any graphs that clearly show an exponential growth in our ability to make sense of the great philosophical questions: "What is life?" "What is consciousness?" Have we seen the law of accelerating returns in our understanding of these questions? Is the growth of wisdom keeping pace with our technology?'

Ray falls silent, contemplating, and then deflects slightly, telling me that he thinks machines are getting closer to what we would call alive and conscious. But I push back. That's not quite the question I asked.

'Well I think these philosophical discussions are now becoming very pressing. For instance, at what point do we have to consider a software program as having human rights? We're thinking more about it than we ever did before.'

It's an echo of something MITs Cynthia Breazeal said to me: that her research into robotics 'pushes us to ask those questions, 'What is life?' 'What is mind?'' She'd told me she thought the answers were 'decades off.' If Ray is right, technology will force our hand far sooner.

I'm staying in Boston for the next part of my re-boot: a meeting with Juan Enriquez, one of the world's leading authorities on the impact of technology (particularly biotechnology) on our economies and politics. Our conversation takes place on the fifteenth floor of Boston's Prudential Tower, in the offices of Excel Venture Management, a venture capital and investment firm Enriquez runs when he's not lecturing around the world, writing books or serving on the boards of numerous businesses and academic organisations.

Juan is a reserved and gentle man in person, reminding me of Kelsey Grammer ('Frasier') although with more hair, a slightly slimmer face and a beard. He describes his life as 'a series of strange accidents,' which is a rather self-effacing way of summarizing an eclectic powerhouse of a CV. Those 'accidents' arguably started rolling off the conveyor belt when, as a young man living in Mexico, Juan walked into his parents' room and said, 'I'm not learning enough here, so I'm going to go to school in the US.'

'I applied late, I had no idea it was hard to get into these places and even though I spoke English – my mother's American – I'd never studied and written in English. I have no idea why I was admitted. I mean during the placement exam I was asked to write a paragraph and I asked, "What's a paragraph?"'

He describes feeling 'utterly stupid' for his first semester but obviously caught up quickly and maintained that accelerated intellectual velocity. He was admitted to Harvard to study government and economics, after which he returned home to 'change Mexico' – a childhood ambition borne out of a belief that his homeland too readily disadvantaged those not in the ruling class. 'I always thought I would work in and change

Mexico. I was bothered by the poverty I saw there.' He became the country's youngest budget director ever, and then returned to Harvard before being offered 'a dream job' back in Mexico as head of the Urban Development Corporation.

So far, so impressive (especially when you consider that during his time in Mexico, Juan was also part of the team that negotiated peace with the Zapatista rebels). And then Juan discovered something more important, a revolution that would not only affect Mexico but the entire world, and all because of some lonely-looking geeky guy at a New Year's Eve party.

He recalls, 'There was this guy sitting over on a corner table by himself and I think "Poor bastard, it's New Year's." I walk over, sit down and talk to him for the rest of the night. By the end of the evening we decided to sail across the Atlantic together in two weeks. By the end of that trip, I had decided that I was going to change my entire career and learn biology.'

The 'poor bastard' in question was Craig Venter, famous for sequencing the human genome. Juan recalls, 'That conversation was so interesting. All of a sudden I thought, "I want to learn about this." I wondered, "Who gets affected by this stuff? What does it do? Why does it matter?"' The sea voyage they took was in search of marine genomes, uncovering myriad new species of micro-organisms. Of the forty-one million genes in public databases forty million come from this single expedition. On the trip Enriquez told *Wired*, 'I gave up a lot for this. I cancelled a meeting with [President] Bush and blew off a couple of foreign ministers.' Well, there's an image. 'But what could be more exciting than sailing around the world, discovering thousands of new species?'

Embarking on the marine voyage, he declared, 'If you do not perceive the possibilities in this shift, if you say *no* instead of *yes*, you will be left in the past. There will be whole societies who end up serving *mai tais* on the beach because they don't understand this.' Juan is very clear that new technologies quickly alter the fate of nations. The rules are changing.

'It took me a damn long time to figure out. It's Darwin. It's the ability to adapt and adopt. It's not the most powerful who survive, it's those who best adapt to change. One of the most important things that people keep forgetting about America and the reason why I think America truly became a world power, is because so many of the founders were adamant about education and science. Just look at Franklin, or Jefferson, and you'll see people deeply committed to critical thinking and education. There was a huge tradition of science and technology education, freedom of inquiry, and that's powered this country in an extraordinary way.'

Thinking back to my conversation with Ray Kurzweil, I give voice to the fear many feel when they hear about the implications of the infotech, biotech and nanotech revolutions currently 'goosing themselves into exponential growth' – that technology is running too fast for us to make sense of.

'If you look at a lot of the things we're building, they're scary as hell to some people. You talk about programming cells or sentient robots or evolving the species using technology – that is profoundly disturbing to some people because this stuff is very powerful. It up-ends industries, it changes how long we live, it changes what our kids may look like. I look at that stuff and say, "Okay, it allows people who couldn't have children to have children, we're going to do away with some of the diseases, and so on." Other people look at that in absolute horror. They say, "Stop the world. This isn't natural. This isn't what God ordered. I want to get off." They're looking for an element of stability and certainty. This desire tends to manifest most during the periods of fastest change, like now. You want something to hold on to. And if you're not part of that ride, if you don't think you can play in that game, then you get an anti-intellectual counterpoint.'

In short, some of us are running scared, which, according to Juan, is about the worst thing you can do. Mark Bedau, editor of MIT's *Artificial Life* and a philosopher with a particular interest in synthetic biology, told me, 'Change will happen and we can either try to influence it in a constructive way, or we can try to stop it from happening, or we can ignore it. Trying to stop it from happening is, I think, futile. Ignoring it seems irresponsible.'

You can stop the advance of knowledge in some places for a while if you're brutally draconian or conservative, but not for long. Beyond that, lost knowledge is soon rediscovered, and the more technology allows autonomy of the individual (from wireless Internet access to the world's knowledge via mobile devices, to power independence through solar technology) the harder it becomes to suppress our advances, good or bad. Or as Septimus Hodge says in Tom Stoppard's *Arcadia* 'You do not suppose, my lady, that if all of Archimedes had been hiding in the great library of Alexandria, we would be at a loss for a corkscrew?'

Juan isn't worried about our self-directed evolution. 'The notion of evolving into something else is terrifying until you consider the question "Are popular radio jocks the be all and end all of evolution?" If that's all Mother Nature wrote, *then* I'm scared. I look at this stuff and say, "If my kids could live two hundred years with a good quality of life, if they could see a lot further than I could, if they could regrow their joints, if they could hear a lot better than I can, if they could have brains that were fifty times as powerful as mine? Good for them. Cool. I'd rather things carry on."'

So, the second part of my re-boot is getting used to the idea that evolution is far from over. It's something I'd kind of understood when I met George Church, and Ray Kurzweil had discussed it too. Juan is ramming it home and he needs to. I've put up a mental barrier to this somewhere in my mind. I have to get my head around the fact that as a *Homo sapiens* I am only a staging post on the way to another form of life that's coming down the line. Attempting to hold on to something 'essentially human' by trying to fight against the very thing humans do best – evolve through culture and technology – is a contradictory and ultimately futile course of action.

But can our morals keep up with all this change? Einstein famously said, 'It has become appallingly obvious that our technology has exceeded our humanity.'

Juan makes an interesting observation.

'Every civilisation has to a greater or lesser extent, some religious moral background. There has to be some evolutionary advantage to having that kind of moral backbone and that kind of belief system, and I think it's because it traces how you move from a hunter-gatherer society, where everybody knows each other and watches each other all day, into a town, into a city, into an empire. ... And just like most animals, almost every religion and God has gone extinct. The interesting question is, which ones survive, how do they survive and how do those moral backbones evolve? We have to ask ourselves, What does a moral ethical framework look like if we start to alter fundamental characteristics of what we consider human?'

The way I interpret this is that if there is an evolutionary advantage to having a moral framework or religion, it had better evolve with you. This, I think, doesn't mean watering down the essential need for compassion, the 'golden rule' of treating others as you would like to be treated, but it does mean we need to work out how to continually keep that ambition central to our societies in a rapidly changing world. There's plenty of work for theologians yet. Indeed, you could argue we've never needed them more, although they, like the rest of us, may have to adjust to a new time frame. Like Ray said, those philosophical questions we've always battled with are going to demand some answers soon.

One thing is for sure. The future won't be a smooth ride.

'Things evolve at different times at different paces, people make different choices and that's one of the reasons countries disappear,' says Juan. 'There really are consequences to your choices. If you choose to shut your doors and not follow technology, you will vaporise your sovereignty. So, there are galactically stupid policies as far as individual countries are concerned. The future of the species worries me a lot less.'

My third stop in search of the re-boot returns me to California to meet a man called John Seely Brown (or 'JSB' as he's often referred to). In a former life he was director of the Palo Alto Research Center (PARC), a hotbed of innovation that is widely attributed with inventing laser printing and graphical interfaces for computers. Now in his late sixties (though he looks a good ten years younger), he works independently, helping organisations adopt and adapt to new technologies. It's on the matter of how our organisations and society will have to change that I've come to talk to him. He dubs himself the 'Chief of Confusion' and sees his job as 'helping people ask the right questions.' He's particularly interested in what he calls 'radical innovation.' His website tells me he is 'part scientist, part artist, and part strategist' as well as a keen motorcyclist. Vint Cerf told me, 'John is among the best thinkers in the United States and probably in the world. He's just a brilliant, modest, very, very deep guy.' After all this, if someone told me he also has a part-time role as Batman, I'm not sure I'd be surprised.

I meet John at his house in the appropriately named district of Professorville, Palo Alto, on a gorgeous California day. As I'd done with Ray and Juan, I briefly recount my trip and tell him I'm looking for some ways of thinking to make sense of it all. He listens intently and I get the impression that every word I'm saying is being catalogued and cross-referenced as I say them. He has the manner of an owl set into the face of a kindly bear and reminds me of my best teachers at school – the ones that always managed to interest you with their subject but never pretended to be friends with you. He imparts knowledge in the same way Keith Richards imparts riffs, with a kind of blistering efficiency and little worry about what this might do to your head.

John believes we're 'moving into a world of constant flux, rather than a world of basically slow, punctuated evolution. If you take this to the extreme, it says nearly every social and technical and business infrastructure we have can't survive.'

Well, we are near the end of the book. What did you expect?

CHAPTER 15

Future shock

When you are through changing, you are through. BRUCE BARTON

The Industrial Revolution changed everything. It changed the way we organised society, the way we educated ourselves, the idea of work and the way we did business. We built an infrastructure the like of which had never been seen before. Railways, roads, sewers, waterways, ports. These were new technologies. Today we don't think of the road or the sewer as a technology. But they are: as Vint Cerf told me, 'If you grow up with a technology, it's not technology. It's just there.'

In 2009, John Seely Brown told a crowd of Silicon Valley business leaders how deep the Industrial Revolution was embedded in their high-tech businesses. What he said isn't immediately obvious but it is profound: 'The structure and architecture of the firm reflects the structure and architecture of the *infrastructure* on which the firms are built,' he said.

In other words, society is built *on top of* our infrastructure, not the other way round: our roads and schools shape us far more than we shape our roads and schools. Or as educator and Abraham Lincoln fan Ken Robinson points out, 'There were no systems of public education around the world before the nineteenth century. They all came into being to meet the needs of industrialism.'

We built our school systems on top of the Industrial Revolution; we didn't get the Industrial Revolution because we'd sent people to school. And the Industrial Revolution is still with us. Cars are not radically different from their counterparts a generation ago, nor are trains, aircraft, roads, railway tracks, airports, the judiciary, schools, universities or our systems of government. 'Our infrastructure has been stable for a shockingly long period of time,' John Seely Brown tells me, 'and we have built institutions on top of it that expect that kind of stability.'

'I've a feeling that's about to change,' I say.

'Right now we are experiencing something we've never experienced in the history of civilisation,' he asserts. 'All past infrastructures have unfolded slowly at first, until they reached a critical mass which then sparked explosive expansion and adoption. Finally things level off and stabilise for decades at a time, often seventy to a hundred years. What's interesting for me is that the infrastructure we're moving into is an exponentially increasing infrastructure, because technology is exponentially increasing.'

'You agree with Ray Kurzweil?'

'I think Ray talks about this to the extreme and he doesn't account for the fact that technology is constrained by society. If you ignore society, it's easy to take the exponential prediction too far. But yes, I agree our digital technology has exponential growth. The question is, how is society going to interact with that?'

By way of example John talks about the introduction of electricity: '[Here] was a new technology, but integrating it with society brought social and organisational innovations too. You couldn't give electricity to the people if you didn't agree on standards and ways to distribute the power, so in the US we started out in a highly fragmented way with each city or county building their own isolated system, but eventually a country-wide standard emerged leading to our nationwide electric grid. That was a huge social and business innovation. All infrastructure is social-technical.'

I'm thinking back to conversations with Rick Hess about the off-grid potential of solar power, how the days of an electricity system dominated by centrally generated power may be coming to an end – and wonder, if it comes to pass, how society will make sense of that.

The current model of electricity generation and distribution is just a tiny example of what John calls 'scalable efficiency' where organisations get bigger in order to drive down costs. These organisations have to be able to predict demand, hoping they are producing the right amount to match it. Too much and their costs will be out of kilter with their income, too little and they miss an opportunity, upsetting the end user. This book is another example of a 'scalable efficiency' operation. Print too many copies and the publisher will be sitting on surplus stock, publish too few and sales are lost.

But all this is changing, believes John. Digital infrastructures allow us to usurp these old ways of working. If you're reading this as an e-book, for instance, then there were almost no physical costs of production and there was no need for the publisher to ponder how many copies of the book to manufacture and ship. They just sent you a file. The reproduction costs are roughly that of sending an email. Everyone wins. As a customer you are never frustrated by a book being out of stock, and the e-version is cheaper too. As a publisher you're never at risk of judging demand wrongly and producing too many or too few copies. The author is happy too, because this reduction in risk and production costs translates into a higher royalty. The book's price becomes a closer reflection of its value to you as an interesting read, rather than as a function of the physical costs of production.

Another option is print-on-demand books, where onsite digital printing and binding machines create a physical book in response to an individual order. Thinking back to my conversation with Eric Drexler, his nanofactory may one day be the ultimate expression of the new digital infrastructure that society is shifting to. Maybe you *are* reading this in the future, having dialled up the manuscript from an online archive that your nanofactory (or whatever home fabrication system you have) made into a physical object. If so, tell your machine I knew its father.

'I don't know if you've sensed it,' says John, 'but innovation is again taking a huge punctuated jump. All of a sudden we are finding completely new ways to create things. These creations are being enabled by the mass of digital infrastructures enabling people to do stuff that would have been impossible before.'

He tells me of a project he's involved with to create a new kind of nanoparticle that mimics the properties of platinum, an increasingly rare resource: 'The sort of thing that only a few years ago you'd have had to ask a national lab to look at and commit many millions of dollars and masses of manpower to. This start-up is doing it with ten young engineers and physicists in a garage where they grab the computational resources they need via the Internet, just buying massive computing power by the hour incredibly cheaply. They're not even starting with combinatorial design [where you try out thousands of compounds to see if you can get something that works] but designing the thing from first principles – i.e., *from the bottom up*. I mean, who the fuck would've tried to go and build a nanoparticle in a garage? It's not something that would have made any sense before. Now it does.'

'Perhaps more importantly,' he continues, 'you now only need a *tiny* amount of money. What we're seeing are very small, very agile companies that are capable of getting off the ground in a way that was previously unthinkable. They don't have an overbearing investor breathing down their backs. They can fund themselves through friends and family.'

I'm reminded of how things are likely to get cheaper as information technology infiltrates manufacturing – the post-capitalist world I'd discussed with Eric Drexler.

John is speaking with a heartfelt urgency now. 'The old-age companies don't know why they have to run faster in order to lose more slowly,' he says, laughing. 'All the practices of those companies are exactly the practices that keep you from being able to engage in the world of fast-paced innovation. They have routines and beliefs built on the assumptions of stability. Almost any company that's more than twenty years old isn't built right for this. In fact, I would argue that companies that are five years old aren't ready either.'

Brown and his co-authors of *The Power of Pull* offer a stark warning: 'The world is transforming around you. The truth is, the thing you did to get there will no longer work to keep you there.' This is something they call 'The Big Shift,' which is already affecting every aspect of society. In particular, educational institutions are grappling to make sense of 'the quickly changing learning needs of students.' As Ken Robinson said in 2006, 'If you think of it, children starting school this year will be retiring in 2065. Nobody has a clue what the world will look like in five years' time. And yet we're meant to be educating them for it.'

I recount something Vicki Buck said to me as we sailed in Queen Charlotte Sound in New Zealand: 'The fear I have is that the education system takes creative, imaginative little kids and seeks to provide conformity and obedience and discipline. I find that scary in a time when those kids are probably going to have to create their own jobs.'

John agrees. 'Education itself is delivered using a factory model. You build up an inventory of skills. You were told these would be useful at some point. What starts to happen when you move into a world where the life of a skill is five years instead of sixty? Our current educational institutions make almost no sense any more.'

And it's not just education. 'Governments are racing to make sense of a world where social and economic changes far outstrip the ability of legislatures and even dictators to keep control,' he says. 'It's a different game now, and many of us have yet to learn how to play it.'

The world I've grown up in is starting to look like a relic, and I realise that almost nothing in *my* education has prepared me for a future built on an exponentially growing information infrastructure instead of a static industrial one. I've been trained to work for organisations that centralise production and I've been spat out of an industrial educational system to fit their needs. What happens as the world shifts toward a network infrastructure and locally manufactured abundance?

Rather than feeling elated by what John is saying, I'm slightly terrified. It's not because I think any of these innovations are a bad idea; it's because quite simply I don't feel ready. It all seems to be coming too fast. I'm almost expecting that by the time I walk out of John's office I'll find the world already transformed into some futuristic society that I've no idea how to operate in. Perhaps I have the first mild onset of a phenomenon identified by sociologist Alvin Toffler – 'future shock' – which can be summarised as too much changing too quickly for you to get your head around.

Part of the re-boot, then, is finding a way to deal with change as the norm and not the exception, realizing that society, having enjoyed a period of relative stability since the Industrial Revolution is, in my lifetime, going to see a radical shift. If the infrastructure we're shifting toward really is exponential in nature, then I need to learn to let go of ideas on a regular basis.

Have you ever agonised about breaking up with someone? You know, that terrible soul-searching period where you know it's not working but wonder whether you've really given them a chance, or put enough effort in, or you worry that the moment they're gone, you'll realise it's madness to be without them? But if it is the right thing to do, you feel sad about the trauma but lighter somehow, like a new window of opportunity has opened up somewhere inside you. You'll take what you learned from the last relationship and make the next one better, you know it.

I think this is how I'm feeling about the industrial capitalist society I've grown up in. I'm scared to break up with it. We've been together so long.

'There may be a whole new system where you're defined more and more by who you are and not by what you own, by what you've created and shared, and what other people have then built on,' says John. 'If you look at what's happening in nanotechnology, in computing, in medicine, you see increasing acceleration of capacity at lower and lower cost – and you're also seeing this trend whereby you are valued not by how much you own but by how much you innovate. The skills that will help you prosper will be based around a questing disposition. That is not typical. That is not considered a twentieth-century skill, but it may be considered more and more a twenty-first-century skill.'

I'm not scared now, but excited. This breakup is starting to feel right.

'We're getting to a place now where we can all live with perhaps a lot less money than we used to, which means you'll see more and more people breaking free of some of the nineteenth- and twentieth-century notions of capitalism as the sole basis of being. But if the world is now accelerating like this, your degree of comfort depends on your ability to embrace change – not flee change.'

'A lot of people and organisations are going to find that difficult, surely?' I say.

'A fundamental clash does exist and the question is, how do you start changing the shape of institutions to embrace this? The real challenge of innovation today is not technological innovation, it's institutional innovation. We have to start inventing new types of institutions that can stay in step with the information age.'

I find this a powerful thought. I've spent my journey looking at technological innovations (from personal genomics through to new ways to fence cattle) but John is saying it will be how we organise our society in response to these ideas that will be the real innovation. We have to break up with the Industrial Revolution and finally form a proper relationship with the Information Age: one based on the continual growth of ideas rather than bank balances.

He concludes our discussion with a story: 'I was asked to suggest the innovation in the last three hundred years that had generated more wealth for mankind than anything else. I knew why I'd been asked – they presupposed I was going to talk about the microprocessor and I disappointed them dramatically. I said, "The innovation that's generated more wealth than anything is the limited liability corporation because that enabled you to accumulate, invest and leverage your wealth but with limited liability." That is what actually unleashed the power of the nineteenth and twentieth centuries, that one innovation. Isn't it curious when you talk about innovation today, you hear *almost nothing* about institutional innovation?'

The third part of the re-boot is complete. Whereas power used to be handed down from above, it's increasingly going to come within reach of us all. Hierarchies are on the way out, networks are on the way in. Don't look to get promoted, look to get distributed – as an individual, as an organisation, as a nation. It's a point neatly made by marketing guru Seth Godin:

> We started with the factory idea. That you could change the whole world if you had an efficient factory that could churn out change. We then went to the TV idea. That said if you had a big enough mouthpiece, if you could get on TV enough times, if you could buy enough ads, you could win. And now we're in this new model of leadership, where the way we make change is not by using money, or power to lever a system, but by leading.

I like the idea that your personal worth in our coming society will be in how many people you inspire, not how much money you have. The rich will no longer be rich because of what they own. The best inheritance you can get, or the most useful thing you can acquire, is what John calls 'the questing disposition.' The value of money, ironically, is decreasing. The value of a good question is rising.

So, what do you want to ask?

The final destination in my search for a re-boot is a set of offices high above the teeming streets of New York. I'm meeting a man who, by his own admission, used to be mildly delusional.

In February 2002, in front of an audience of some of the world's leading thinkers, he said, 'I think that I may have believed unconsciously then that I was kind of a business hero. I had this company that I'd spent fifteen years building ... It had recently gone public, and the market said that it was apparently worth two billion dollars, a number I didn't really understand.'

But then it all went horribly wrong as the dot-com bubble burst. 'I entered eighteen months of business hell. I watched everything that I'd built crumbling ... [It] took eight years of blood, sweat, and tears to reach three hundred and fifty employees ... In one day we laid off three hundred and fifty people, and before the bloodshed was finished, a thousand people had lost their jobs from my companies. I felt sick. I watched my own net worth falling by about a million dollars a day, every day, for eighteen months. And, worse than that, far worse than that, my sense of self-worth was kind of evaporating. I was going around with this big sign on my forehead: "Loser." And I think what disgusts me more than anything, looking back, is how the hell did I let my personal happiness get so tied up with this business thing?'

Today Chris Anderson is a whole lot happier. He managed to save his magazine publishing business but then sold it on and for the first time in years had time on his hands. He started reading again, 'discovering, *holy crap!* There's so much amazing new thinking out there. That was a kick.' This revelation led him to buy and become the curator of the TED (Technology, Entertainment, Design) talks. He'd lost interest in money as a currency and had become interested in the currency of ideas. He had rediscovered what John Seely Brown calls 'the questing disposition.'

TED, Chris tells me as I take a seat, is a conference platform for 'people who can offer a lens through which to see the world in a different way.' He speaks with almost boyish enthusiasm, which matches his looks. He's just past fifty but seems ageless. From a distance he could easily pass for early-twenties. There's an openness to him that is reflected in his posture, the soft tones of his voice, the way he's happy to be approached by ideas and happy to approach them. He's thoughtful without being serious, sharp without being aggressive, friendly with no need to make friends. My first impression is almost that of a blank canvas, but as our talk progresses I change my mind. He's not a blank canvas, he's a constantly changing one, repainting himself willingly when new knowledge comes his way.

Every year, Chris and his team gather together some of the world's best disenthrallers and give them eighteen minutes apiece to tell the rest of the world how they see things. In 2006, the talks started going on online and hundreds of millions of downloads pay testament to a hunger for new ways of looking at the world. TED does no advertising, it's a word-of-mouth thing and it's easy to see why. TED ideas are by turns so mind-bending, hopeful, scary and entertaining that they demand to be shared.

Mixing the most radical ideas with a short format means speakers need to hone their presentations, and because the Internet audience is millions there is no room for academic long-windedness. In 2007, statistician Hans Rosling (whose thoughts on population I'd considered while walking the packed streets of Malé) delivered a blistering attack on the concept of 'developed' versus 'developing' nations ... and then he swallowed a sword, because sword swallowing is 'a cultural expression that for thousands of years has inspired human beings to think beyond the obvious.' Pure TED.

On TED.com, too, you can see Juan Enriquez talk about the coming age of genomics, Ray Kurzweil summarise his law of accelerating returns, and Hod Lipson demonstrate his self-aware robots. You can see Ken Robinson ask us to re-evaluate what we need from our educational systems, Steven Pinker tell you the world is less violent and then watch Robert Wright explain why that might be, along with a host of other mind-shifting presentations that make you see things from a different angle.

I've come to see Chris because in putting together the TED talks he is probably assailed by more new ideas than pretty much anyone on the planet – and he has to make sense of them somehow. I ask him how he synthesises everything.

'Well, like everyone else I'm on a journey,' he says. 'But there's a very boring view of the world which is that 'things happen' and basically history is one thing after another, the idea that you really can't say much about the future, other than it's probably not going to be as good as you think. But there are bigger trends at work, like Ray Kurzweil's "law of accelerating returns": the idea that it's in the nature of ideas themselves to build on one another.'

Like John Seely Brown, Chris has seen a quickening of the pace of innovation in recent years. 'The acceleration of knowledge and ideas made possible by the fact that humanity is connected for the first time is vast,' he says. 'This is happening in hundreds of areas of human endeavour.'

By example he talks about how the quality of contributions to TED has been improving since the conference went online. 'Speakers are looking at what other speakers are doing and are putting in far more preparation time than they ever used to.' (How many, I wonder, are taking sword-swallowing lessons?) 'It's a big kick for me to see these people competing. I've started to call it "crowd accelerated innovation" and it's not something exclusive to TED.'

Chris thinks that rather than technology separating us from our humanity, it could help us to rediscover it. After all, what could be more human than the Legion of Extraordinary Dancers, who had come up in my conversation with Vint Cerf – a dance troupe built from kids innovating on each other's moves (continually posted on YouTube) and who appeared at the 2010 TED conference in Long Beach, California.

'We're pretty sure the biggest growth in the TED audience is people in their teens and twenties,' says Chris. 'A lot of people fear that the generation coming in are made up of non-curious and not very hard-working slackers, but there's a lot of evidence contrary to that. Increasing interconnectedness is changing the psychology of the generation coming through; they really believe that they can make a difference to the world.'

'So it's not just that they're curious. They also have a sense they can do something?'

'Exactly. I think the sense of your ability to play a role in an interconnected world is changing. And the reason they believe that is because it's true.'

'Does that make you optimistic?' I ask.

'I don't know that the future's going to be better,' says Chris. 'But I think there's a very good chance that it will be and I think there is something everyone can do to further increase that chance. Of course, the future might well be truly horrible. I think it's all to play for and I think everyone of sound mind and conscience should be in the game, trying to shape it in the right way.'

As I'm leaving, Chris asks about my solar bag. I explain the technology and how Konarka was founded on the principle of taking cheap power to areas without electricity grids.

'That's superb!' he exclaims. 'Fantastic. Fabulous. There are so many things like that that can get you really excited.'

At the door he says, 'You know the title for the next TED conference?'

'No,' I reply.

'We've decided to call it: "And now the good news."'

I take a walk through Manhattan, and find myself sitting on the steps of the New York Public Library watching the crowds go by. It's a baking hot day. Did I get what I came for, the final part of my re-boot? There's something in my conversation with Chris that's bashing around inside my head but hasn't settled.

Often when I see crowds of people I'm struck by the fact that inside every single head there is a separate personality, a separate consciousness, a unique story. I sometimes try to imagine what that story might be. But today I find myself thinking about our collective story and who tells it. We all have our individual tales but who looks after the story of humanity? We have plenty of biographers (they're called historians), but who is scriptwriting the next episode?

I'm thinking about the world I've grown up in, recalling the story of the future I've been used to hearing in the discourse of politics and its reflection in the media. Despite everything I've seen on my journey, I feel my heart sinking. My perception of the story we seem to be telling ourselves is: 'Life happens to you, the future is not going to be very good (especially if you vote for that guy), it was better in the old days, you've got to look after yourself, the world is violent and unsafe, your job is at risk, the generation below you are feral and dangerous, things are changing too fast and you can't trust anyone.'

What's worse is that it's a story that seems pretty constant. I can't remember a time without it and, as I think about it, I can feel its fingers reach deep into my soul. I find myself chastising the optimism of those I've been visiting, and my own naïveté in being taken along with it.

And then the final part of the re-boot goes off in my head like a firework.

CHAPTER 16

Genomed

Following the light of the sun, we left the Old World. CHRISTOPHER COLUMBUS

I've travelled over sixty thousand miles across four continents, talked to over thirty geniuses, met four robots and had two terrible conversations with computers. I've contemplated immortality, the end of capitalism and a new age of human evolution. In the process, I've attended an underwater cabinet meeting, helped invent a cocktail, seen spaceships, been insulted in the Outback, made a brace of new friends and had one near-death experience (thank you, Tracy). I'm not the same person I was when I started.

Now I'm sitting in my local park overlooking London. Just behind me is the spot where in 1795 the British Admiralty erected its optical telegraph station to pass signals down the line between coast and capital. Communications have come a long way since 1795. On my lap is a computer, battered and grubby from long hours on the road. Using my mobile phone as a wireless modem I am surfing the Internet. In particular I am looking at my 'genetic profile' having just logged on to the website of 23andMe, the Google-funded personal genomics company that Sergey Brin has been using for his Parkinson's research.

Several weeks ago, the company sent me a plastic tube, which I filled with saliva and returned to its laboratories. From this the company extracted cheek cells, out of which they stripped my DNA to be duplicated many times over. These synthetic copies of my DNA were then chopped up and applied to a 'DNA chip,' a glass slide with millions of DNA 'probes' on its surface.

The DNA probes are mirror images of locations on the human genome that are known sites of genetic pointers that can indicate your susceptibility to particular diseases, the chances of you exhibiting a particular trait or give clues to your ancestry. Because DNA's component parts love to snap together in predetermined ways, copies of my own genetic code would have latched on to these probes with varying degrees of connectedness, letting the company know which markers I possess and allowing them to suggest what these might mean for my health. Because our understanding of the interplay of our genes and environment is still evolving, 23andMe attaches confidence ratings to each finding (the higher the rating, the more secure they feel in their analysis).

Because I have one genetic marker that a 2007 German study suggests is linked to Tourette's syndrome, 23andMe let me know I *might* have an elevated chance of the condition, although they give this a confidence rating of one (out of four). In the two-star category there are potential elevated risks of 'essential tremor,' 'Hashimoto's thyroiditis' and 'Sjögren's syndrome'. The company gives a confidence rating of three to its analysis that I have higher-than-average risks of asthma, atopic dermatitis and chronic lymphocytic leukemia.

This is all starting to sound depressing until I dig deeper and discover that the average population's risk of getting these diseases is small in the first place and that my 'above average' susceptibility is still a low risk overall. In addition, a lot of the research linking particular genes to medical conditions is preliminary and therefore not yet conclusive.

And just as I have elevated risks of some conditions, I have reduced risks of others, including lung cancer and melanoma. Apparently I am less likely to be addicted to nicotine than your average Joe, but more likely to get hooked on heroin should I give it a try. So, guess I'll give that a miss then.

I pay more attention to the risks given four stars (although these still come with caveats). It seems my rear end is the weak point. Apparently I am 1.47 times more likely to get prostate cancer than your average bloke, and 1.43 times more likely to get colorectal cancer too. I'm not that bothered. Knowledge is power and I can now try to proactively reduce my risk of both conditions through some simple lifestyle choices. For instance, eating fish has been linked to reducing the risk of prostate cancer and I like fish, so bring on the (responsibly farmed) salmon. I'll make a point of getting screened regularly to detect anything untoward early on as well.

I'm actually less struck by the information I'm reading than by how it came to me. I'm looking at genomic data (reminding me of my trip to see George Church at Harvard Medical School) that has been provided by a piece of biological nanotechnology (bringing to mind a coffee shop talk in Palo Alto with Eric Drexler). On the grass next to me is the bag given to me by Konarka. Its organic flexible solar panel (that rolled off that 'big printer' in New Bedford) has been charging the mobile phone I am now using to access the Internet (co-fathered by Vint Cerf, whom I met in Washington, D.C). Thinking back to my conversation with Ray Kurzweil about exponentials, I realise that all of this technology would have cost millions of dollars ten years ago. Somewhere on my computer is Hod Lipson's 'truth finding' software brain that I marvelled at in Ithaca. Given the right data, my humble laptop is now capable of discovering new knowledge.

The sun pokes through the clouds and my thoughts move to what I've learned about our battle with climate change. Looking into the sky I remember Klaus Lackner's machines that can take CO_2 out of the air. I remember how that CO_2 can be used as a raw material to create liquid fuels that could replace gasoline, harnessing the power of synthetic biology. Maybe such fuels will one day power spacecraft like those I saw in Mojave? I think of Vicki Buck and her algae fixation and the Black Phantom that takes waste biomass and turns it into useful products, including carbon-sequestering charcoal. The grass between my fingers brings to mind my journey through the Australian bush with Bruce and Tony and those fence lines that separated areas of growth from bare soils, signalling a system of agriculture that is rejuvenating biodiversity and profits on Australian farms simultaneously, all the while sequestering carbon into the soil.

As I sit on this hill writing these words, I know that what I have seen is only the tiniest fragment of the innovation going on and an even tinier fraction of the innovation yet to arrive. I know that advances are coming faster than Paralympic athlete Oscar Pistorius, whose carbon fibre blades give athletics officials the willies about his 'unfair advantage.' I have to make peace with the fact that this book is already a historical document. It's less a posed portrait, more a blurred snapshot of a rapidly innovating world. An action shot, if you like.

In his book *Engines of Creation* nanotechnologist Eric Drexler approaches the future by asking three questions: What is possible? What is achievable? What is desirable? The first question seems easy to answer. As we learn to control the very atoms of matter, the mechanisms of biology and the power of computation, there is in fact very little that we can't do in a physical (and indeed virtual) sense. Solutions to climate change? Already developed. An end to the energy crisis? No sweat, sign on the line. Increased food production? Why didn't you ask? Holidays in space? Join our frequent flyer program. Cheaper, better manufacturing through nanotechnology?

We're working on it.

But when we ask what is *achievable,* well that's a different story. Because what we achieve will largely be determined by what we collectively decide is desirable. As George Church told me all those months ago at Harvard Medical School while we discussed personal genomics,'The only thing that puts this kind of medicine far away is really will, right? The question is, how motivated are we?'

A lack of will isn't our only failing, though. Sometimes we're too eager to accept the easy promise of technology. History is replete with techno-optimists who claimed the next revolution from science would usher in a golden age. The most bombastic of their number believe there is no problem technology cannot solve. Two years after the optical telegraph station that once stood behind me was erected, the *Encyclopaedia Britannica* boldly suggested the networks of which the station was part would soon settle disputes between nations in hours instead of years. Over two hundred million people have died in conflict since.

That said, I've seen so much on my trip that has made me feel optimistic. It's hard not to smile when you see what George Church is doing to the world of medicine. He's providing tools that have the potential to save millions of lives (although as a scientific adviser to 23andMe, he's also partially responsible for the death of a few extra fish, now I'm going to be eating more). It's difficult not to feel a glimmer of hope when you meet Klaus Lackner, or see Carbonscape's Black Phantom, or realise the potential of nanotechnology to underpin revolutions in energy production, health care and sanitation. The other day my editor asked me to list some of the things I'd discovered on my journey, to get their sales team enthused about the book. Off the top of my head I wrote:

> Replacement body parts grown from patients' own stem cells are already here.
>
> Bionics could soon be able to help the 'handicapped' outperform the rest of us.
>
> The scene in *The Empire Strikes Back* where Luke Skywalker gets a new hand has already happened for real.
>
> We're learning to recode cells – curing disease from the inside.
>
> We're all living longer, and globally the gap between rich and poor, healthwise, is narrowing.
>
> The genomics revolution means drugs we can't use today could be released to the market (and drugs should be cheaper).

We can stop pandemics (even those made by bioterrorists).

There's no shortage of 'fossil fuels,' we just won't make them out of fossils (and they'll be carbon neutral).

Solar power is finally coming. There's an 'off-grid' revolution starting up.

We're harnessing the power of life to make everything from drugs to plastics – also known as 'How bacteria are being used to make things cheaply.'

Nanotechnology could change everything, including ending capitalism, and ushering in an age of abundance.

Machines will become friendly (the cutting edge of robotics looks nothing like *The Terminator*).

We can take CO_2 out of the air, reduce carbon emissions and increase food production (while reducing its cost).

We're heading back to space.

The Internet is a laboratory for human good . . .

But there's one thing technology cannot solve. A lack of imagination. How is it that we have not managed to equitably distribute the wealth we create? There are millions of people who do not feel the same dividend of increased life spans and better health care that industrial capitalism has brought to some. One in five child deaths is the result of diarrhoea. The same proportion dies of malaria. (In Africa, a child dies every forty-five seconds from the disease.) The Rwandan genocide was estimated to have killed eight hundred thousand people in just a hundred days. The predominant technologies on show there were weapons of slaughter. Those people died in the same world where someone thinks it's a good idea to manufacture a lady's handbag that costs a quarter of a million dollars.

Should I do the whole list? The catalogue of injustice, violence, prejudice and inequality that still besets us? I could fill this book again if I did. But as far as books go, that market is pretty much covered. In *The Rational Optimist*, Matt Ridley recalls browsing the current affairs section of an airport bookshop and, despite volumes on offer from authors as diverse a Noam Chomsky, Barbara Ehrenreich, Al Franken, Al Gore, John Gray, Naomi Klein, George Monbiot and Michael Moore, not seeing a single optimistic volume. 'All argued to a greater or lesser degree that (a) the world is a terrible place (b) it's getting worse (c) it's mostly the fault of commerce and (d) a turning point has been reached,' he lamented.

I think Ridley is being a bit harsh on those books. I interpret them as heartfelt calls to action: dire warnings followed by suggestions about what to do to make things better. You don't have to agree with everything, or indeed anything they say, but if it is doom-mongering it is at least *pro-active* doom-mongering, motivated by the belief that things can get better or that the worst can be avoided. Concerned friends don't always tell you what you want to hear.

My good friend Eliza says, 'The vast majority of the world's natural resources are controlled and divvied up by old men sending young men to war to protect their empires of traditional power, and that has not yet changed much.' Eliza's great at this kind of thing. She's just the sort of person to remind me that some histories of Christopher Columbus, whose quote I use to open this final chapter, mention his habit of murdering natives (and I value her for it). She really *could* have filled this book with all the things that are, and have been, wrong with the world. She's a walking encyclopaedia of doom who doesn't see the contradiction in being pessimistic and progressive. 'Why can't you be pessimistic but remain committed to all progressive options?' she asks. 'I have an issue with the idea that we should all be optimistic, it seems to be disrespectful to people who are suffering and dying now because of our messed-up world.'

Sitting in Telegraph Hill Park, London, I've come back to the thoughts that first occurred to me sitting on the steps of the New York Public Library – that any dream of the future must be shaped and influenced by the society in which it is dreamed. I do not aspire to the same future that my cave dwelling ancestors did. Ray Kurzweil remarked, 'If you asked people ten thousand years ago what would solve the problems of the world, they'd say, "If we can find a way to move bigger stones in front of our caves to keep the animals out and a way to keep these fires going, that would be utopia and we'd solve all our problems." It's back to what John Seely Brown and I had discussed – how the frame shapes the picture more than we comprehend.

For all the marvels I've seen on my travels, I realise that what has impressed me the most is not the innovations on display but the people I have met. Because nearly all of them seem driven by one question: How can I make the world better? This question has framed their lives more than anything else. And you know what? Nearly all of the people who I feel are asking themselves that question seem really happy.

My personal manifesto for approaching the future came together on the steps of the New York Public Library and I'm feeling comfortable with it now. I don't expect anyone else to follow it, but I'm going to give it a go. I will endeavour to embrace the exponential, come to terms with the fact that evolution isn't over, understand that institutions that don't innovate in the way they organise themselves (including the ones I work for) might have office space going soon, adopt a questing disposition and measure my value by what I create, not what I own.

The future is going to be a rocky ride. Many new technologies will have their Three Mile Islands and be co-opted in the service of more 9/11s. People will still fight. Injustices will not disappear. Andrew Lloyd Webber musicals will still be staged. But now that my journey is complete, I steadfastly refuse to believe that human society can't grow, improve and learn; that it can't embrace change and remake the world better, that we can't make the underlying trend upward. The musician Nick Cave summed it up, I think: 'People think I'm a miserable sod but it's only because I get asked such bloody miserable questions,' he said.

Here's a question that is neither miserable, nor cheerful. It's the one I started out with.

'So what's next?'

I asked it because I wondered what my life might look like. What I've discovered is that my question is its own answer. There is a new story of humanity waiting to be told if we respond well to its challenge. I reckon we can have a go, don't you think?

Right, I'm off to eat some fish.

FURTHER READING AND NOTES

The further reading and website selections below are a personal choice of sources that I found most useful and interesting in researching and writing this book. They are laid out by section but, of course, many apply to multiple areas.

Notes and references for each citation in the book have been posted online, as the majority of them are web links. They are laid out by chapter and page at:

www.anoptimiststourofthefuture.com

MAN

Juan Enriquez, *As The Future Catches You: How Genomics & Other Forces Are Changing Your Life, Work, Health & Wealth*
Crown Business, New York, 2001
A truly excellent summary of where 'the life sciences' are taking us. Juan told me this book started off at 3,000 pages – knowledge he spent six years condensing into less than an evening's read, after which you'll see the world differently.

MACHINE

Rodney Brooks, *Robot: The Future of Flesh and Machines*
Allen Lane, New York, 2002
The history and the future of robots from one of the discipline's most inventive and challenging voices. Brooks has a light and readable style with many enjoyable anecdotes woven in.

K. Eric Drexler, *Engines of Creation: The Coming Era of Nanotechnology*
Anchor Books, New York, 1986
Nanotechnology starts here. The fact that this book is still selling briskly as it approaches its twenty-fifth anniversary is testament to the power of the ideas within, and indicates how far nanotechnology still has to go. Arguably the most important idea of the twentieth century.

Tom Standage, *The Victorian Internet: The Remarkable Story of the Telegraph and The Nineteenth Century's Online Pioneers*
Weidenfeld and Nicolson, London, 1998
Vint Cerf called this 'a great book' – and he's right. A history of the electric telegraph that puts our increasing connectedness into perspective.

EARTH

Chris Goodall, *Ten Technologies to Save the Planet*
Profile Books, London, 2008
A highly readable account of 'renewable' technologies that cuts through the hype to deliver a measured picture of how we might green our economies.

Wally Broecker and Robert Kunzig, *Fixing Climate*
Profile Books, London, 2008
Perhaps the definitive history of climate science for the layman – and a powerful take on the future of geo-engineering.

Allan Savory with Jody Butterfield, *Holistic Management: A New Framework for Decision Making*
Island Press, Washington DC, 1999
A farming manual – but one that'll change the way you make decisions.

Stewart Brand, *Whole Earth Discipline*
Viking Press, New York, 2009
This is the book that managed to cheer up James Lovelock. Brand told me 'my views are strongly stated and loosely held' – he wants a debate, and there is plenty in here to start one. Lucid, powerful and, despite its breadth and depth, a surprisingly easy read.

RE-BOOT

Ray Kurzweil, *The Singularity Is Near*
Penguin, New York, 2005
Douglas Hofstadter pronounced the ideas in this book a blend of 'very good food and some dog excrement'. That's unfair but gives you some indication of how radical Kurzweil is. You may not want to agree with everything he

says, but the argument over what we'll do with technology is one of the defining battlegrounds of this century and it's mapped out nicely here.

John Hagel III, John Seely Brown, Lang Davison, *The Power of Pull*
Basic Books, Philadelphia, 2010
A business book – but a wake-up call for anyone who still thinks the world is a hierarchy.

Websites

www.ted.com is one of two emerging de facto places to get a grip on future possibilities. The other is **www.edge.org**, the online presence of John Brockman's Edge Foundation. Both come highly recommended, but be warned, you might lose a few days.

Also, pop over to **www.gapminder.com** and let Hans Rosling show you how to make statistics interesting and change the way you understand the world as a result.

ACKNOWLEDGEMENTS

My heartfelt thanks to: Caroline Smith, Adrian Mukasa, Amy Stevenson, Andrew Billen, Andrew Mosely, Andy Ross, Andy Smith, Anke Hofmeister, Anthony McCann, Arthur Gleckner, Bill Mitchell, Bob Henson, Brian McKenzie, Bridget McKenzie, Bruce Ward, Charlie Viney, Chris Anderson, Chris Goodall, Colin O'Carroll, Cynthia Breazeal, Dan Reicher, Dan Stiehl, Darren Sanders, Dick Rutan, Duncan Brown, Eliza Hilton, Emmanuel Skordalakes, Eric Drexler, Eva Shivdasani, Felicia Spagnoli, George Church, Grahame Strong, Greg Atkins, Hannah Williams, Harald Willenbrock, Hod Lipson, Holger Emrich, Howard Berke, Imran Yasin, Jack Milner, Jeff Greason, Jim Aach, John Seely Brown, John Ward, Jon Collins, Juan Enriquez, Karen Carney and boyfriend Bruce, Karen Wright, Kat Arney, Katherine Rose, Ken Haylan, Klaus Lackner, Kristine Kelly, Larry Weldon, Laura Galloway, Lisa Goldberg, Lucy Billen, Malcolm Bacchus, Marc Zaballa-Camprubí, Mark Bedau, Mark Ellingham, Mark Philips, Martin Birchall, Megan Mosely, Meredith McDonough, Michael Coughlan, Mitchell Burns, Mohamed Nasheed, Mohamed Ziyad, Natalie Spanier, Nick Bostrom, Nick Gerritsen, Nikki Roswell, Paul Milnes, Paul Roberts, Penny Daniel, Per Jensnaes, Peter Dyer, Peter Norvig, Peter Rigaud, Peter Schöni, Philip Groom, Phillip Stark, Polly Guggenheim, Rachel Holtzman, Ray Kurzweil, Rebecca Gray, Rebecca Suffling, Renate Ruge Christine Axton, Rene Hen, Richard Jones, Rick Hess, Rick Jenkins, Rob Fenwick, Robert Kunzig, Rodney Brooks, Ruth Killick, Sam Tobin, Sandra Higgison, Sarah Reed, Sarah Ward, Sonu Shivdasani, Sophy Williams, Stephen Aguilar-Milan, Stewart Brand, Stuart Witt, Taragh Bissett, Tim Langley, Tim Tully, Tim Wright, Tony Kensington, Tony Lovell, Tracy Wemett, Vicki Buck, Vincent Heeringa, Vint Cerf, Wally Broecker, Xavier Claramunt and Yveta Marsarova.

EPILOGUE – A YEAR ON

What happened next ...

Never give up hope, you know? Never give up. Just keep moving. Tomorrow must be better. Tomorrow is better. MOHAMMED NASHEED

A crowded room below a pub in the city of London. I'm about to address the assembled drinkers and am somewhat terrified. Given I've become used to public speaking, this may seem odd. Since the publication of *An Optimist's Tour of the Future* in 2011, I've been all round the world talking to, and working with, all manner of people – from schoolchildren to the boards of multinationals – to consider philosophies, principles and techniques for approaching the future. So experience, coupled with having done time as a stand-up comedian, means I'm generally not as apprehensive as many people are about taking a stage. (One exception was being asked to address a private meeting of the world's super-wealthy. 'I don't want you to be nervous,' said one of the organisers just before I went on, 'but this audience represents $250 billion.' There were less than a hundred people in front of me. Also, I was about to tell them something I wasn't sure would go down all that well ... of which more later).

Today is different though. These people haven't asked me to talk. Instead I've asked *them* to convene to see if a crazy idea might work, an idea born out of hundreds of conversations I've had both in person and on-line with readers of the book, and people who've attended talks. I've had versions of this same conversation with accountants, taxi drivers, teachers, CEOs, waitresses, doctors, plasterers, even a priest, and it goes something like this:

'Hey Mark, I really enjoyed your book.'
'Thanks!'
'But it's given me a problem.'
'Oh, what's that?'
'Well, now I can see there is the possibility of a better future I want to help make it happen. So, can I ask you something?'
'Sure, go ahead.'
'What should I *do*?'
'Do?'
'Yes.'
'About what?'
'The future.'

At first I demurred. I couldn't tell anyone what to do. That would be arrogant.

At which point some people would get annoyed. 'Well, it's alright for you,' they'd say. 'You're an author, swanning around the planet meeting all these cool people. You might be able to do something because you're connected. You can influence people, but I'm a [insert job title here]. What do you expect me to do? I've got to look after my family.'

It seemed that for some people the book had given them a glimpse of a (possibly) better future, but one that they felt they had no role in shaping. After all, very few of us are world-leading scientists like George Church, Cynthia Breazeal or Klaus Lackner. As the economic crisis continued to rage around us, the gap between the interviewees in my book and many readers I was meeting felt too great. The future might still be worth playing for but I was increasingly realising that to many people it appeared that it was a game for a select few.

At the root of many people's frustrations are experiences of trying to make a positive change and being foiled by their employers. I've lost count of the stories I've been told of proposed innovations to improve things being thwarted by a prevailing culture within organisations. Everywhere I go now I am constantly reminded of my meeting with John Seely Brown, and our discussion about how the stamp of the Industrial Revolution looms large over the way we continue to organise ourselves – a system built on the premise of specialisation, where people are purposely separated from each other so they can 'focus' and gain ever more detailed experience of their allotted task. Yet innovation almost by definition occurs when two ideas that haven't previously met are allowed to interact – 'ideas having sex', as Matt Ridley pithily puts it.

One of the great ironies of today is that so many organisations *say* they want to innovate, to become greener, more responsible, more creative blah, blah, blah ... but when it comes down to it they seem incapable. They find it almost impossible to free up their employees to smash ideas together in the cause of finding new and better ways of doing things. This isn't just corporations. It applies equally to most schools, universities, government departments and NGOs.

That's why when I'm called into an organisation I have a test that tells me early on who genuinely wants to create an enabling environment for their staff and who just wants some 'innovation-wash'. I look the management in the eye and ask, 'OK, so how many rules can we break?' On rare occasions the big boss says, 'All that are necessary, within the law,' or something similar. More often than not, however, there's an awkward pause while everyone looks at each other as if seeking permission, then their shoulders sink slightly as if something is quietly dying inside them. It's at this point someone will say something along the lines of, 'Well, we have these processes and procedures that are very important to us, that embody the spirit of the organisation that we've spent a long time developing and embedding and so, you know, we'd really like to innovate within that framework...'

As John Seely Brown impressed upon me, our workplaces and schools shape us far more than we shape our schools and workplaces. It's no wonder so many of us feel stifled when it comes to shaping a future different from our past.

And so, here I am in The Red Herring pub, Gresham Street, London EC2, about to launch what I hope will be at least one countermeasure to all that stuck-ness and frustration so many people were telling me they were feeling.

'Welcome to the first meeting of The League of Pragmatic Optimists,' I say.

I'd been inspired by how the TED talks have become a phenomenon – reaching millions of people with 'Ideas Worth Spreading'. But I'd heard the same criticism levelled at TED that people were levelling at me. It's alright for those with MIT professorships and publishing deals to bang on about making the world better, being given the platform of a conference or a book deal to espouse their observations or viewpoint, but what about the rest of us? I wanted to give an answer to the question 'What should I do?' and the answer was to try and create a space where people could unleash their employer-stifled creativity and ambition, where frustrated do-ers could meet with like-minded individuals, get tips from each other on how to change the system, or get a shot in the arm after another month of battling the beast; where accountants, taxi drivers, teachers, CEOs, waitresses, doctors, plasterers, priests and anybody else could smash their ideas together and find some projects to get involved in, a place for 'Ideas Worth Doing'.

Tentatively I mooted the idea on Twitter, Facebook and to those who'd subscribed to my email list having read the book. We advertised sixty-five places (capacity for the venue) and they were snapped up in less than an hour. Bloody hell.

I am convinced there is a hunger for getting involved in the narrative of the future. At that first meeting five people were brave enough to stand up and talk about something that they thought could be improved and they wanted help with. The projects ranged from protecting the rights of African musicians, to creating a series of workshops to help unemployed graduates start their own businesses, to a project using neuro-feedback (electroencephalography) headsets to teach students 'mindfulness'. All of them got help, one even got funding. It was all going so well! But towards the end of the evening a chap called Chris approached me, and as we spoke my heart sank.

'Hey Mark, I really enjoyed tonight, but I have this problem.'

'What's that?'

'Well, I'm not like these people here. I can't see how I can help anyone.'

'You want to know what you can do?'

'Yeah.'

The evening had failed him.

'Well, what do you do for a living?' I asked.

'I'm a tax accountant.'

I'll admit it. I had nothing. And then he said, 'I suppose that I could maybe give free tax advice to any projects, making sure they get more cash into whatever they are doing and give less to the taxman. Would that help?'

I put this to the meeting and Chris got a massive round of applause, something I'll warrant he's never had in the office.

Bingo.

A few months later we held another meeting where the projects proposed included helping underprivileged kids get into media internships usually only open to graduates with parents who could afford to subsidise them (and the proposer went away smiling with a bag of new ideas). At the third meeting (Chris was back at this one) we had a diverse set of proposals once more, ranging from someone wanting help in putting together an event on climate change ahead of the forthcoming Rio summit, to a project hoping to create a new walking route through London (this got lots of people excited, not least the originator who couldn't believe how much help was offered to him). It seems that by the time you have forty or more people in a room somebody knows something about nearly any subject (or knows someone else they can put you in touch with).

By now we had people wanting to set up chapters of the league in Brighton, Geneva, Glasgow, Madrid, Oxford, Singapore, Sydney, Boston and New York. It seemed that the league was good if you had something specific you wanted to do, but was also good if you just wanted to get involved in doing something good. Our problem now was finding the resources to support the groundswell of interest.

Simultaneously of course, all those amazing people I had written about in the book had hardly rested on their laurels.

Tony Lovell has secured hundreds of million of dollars from pension funds to further his project re-greening the outback.

A team led by the Wellcome Trust Sanger Institute in Cambridge has successfully removed stem cells from a patient with a genetic liver complaint, and used a gene therapy to *edit out* the mutation responsible for that condition, replacing it with a functioning gene sequence. These edited stem cells were then grown into healthy liver cells. In short, the genetic code responsible for the disease had been rubbed out of the patient's genome and all of sudden the previously incurable looked curable.

In May 2011 I found myself giving a talk at IBM and stood next to Watson, a computer capable of answering questions posed in natural language (Watson wowed the world by beating expert champions of the American game-show *Jeopardy!*)

Almost daily it seemed nanotechnology breakthroughs were hitting my inbox, many focusing on the ability of nano-engineered medicines to target specific cells.

In October 2011 I shared a stage at Austria's annual Elevate Conference with Google's Vint Cerf (admittedly he was there virtually via Skype) where the role of the Internet in facilitating the then-burgeoning 'Arab Spring' was discussed, recalling our conversations about how increasing interconnectedness drives positive social reform.

Klaus Lackner became a finalist in the Virgin Earth Challenge (Richard Branson's $25million prize for 'whoever can demonstrate to the judges' satisfaction a commercially viable design which results in the net removal of anthropogenic, atmospheric greenhouse gases'). In a meeting of all eleven finalists I was pleased to find out that he was already in conversation with Sapphire Energy who feed CO_2 to an algae they've engineered to make crude oil. Rather than ship the CO_2 to their algal ponds in trucks as they currently do, they're hoping to take it from the atmosphere, creating a carbon-neutral fuel source.

As the months went on from publication (and the book was released in more countries) I found myself increasingly hanging out with the sort of people I'd interviewed, and began working with some of them too. Towards the end of *An Optimist's Tour...* I'd concluded that what impressed me most was 'not the innovations on display but the people I have met.' All of the innovators I met seemed to be driven by a fundamental question: 'how can I make the world better?' But I began to see it was more than that, realising there were other core principles embedded in the minds of these extraordinary do-ers.

As part of establishing the League of Pragmatic Optimists I've begun to codify these. I offer them here because I've found them incredibly useful ways of thinking when you're trying to turn an idea into a reality.

Have an unashamed optimism of ambition. Don't feel embarrassed to say that things can be better. Have no qualms about imagining an improved world and advocating for it, no matter how much derision you may receive at the hands of the cynical.

Engage in projects that are bigger than you are. The philosopher Daniel Dennett says that an occupational hazard of his profession is being asked 'what's happiness?' The best definition he's come up with is 'Find something more important than you are and dedicate your life to it.'

Engineer Serendipity. Get out of your comfort zone. Find ways to come into contact with new ideas and people whenever you can. Pragmatic optimists happily let their ideas go dating.

Making mistakes is OK, but not trying is irresponsible. As Ken Robinson says, being wrong is not the same as being creative, 'but if you're not prepared to be wrong you'll never come up with anything original.' As Edison pointed out, he didn't come up with the right way to make a light-bulb without first coming up with 3,000 wrong ways to make a light-bulb.

Whether we like it or not we are defined by what we do, not by what we intend to do. Pragmatic optimists aren't interested in what you *might* do if you had more time, or if your manager was more understanding, or if *you* were the manager, or if it was next week. You are what you do. That's it. Get on with it.

Ask for evidence. Stories and anecdotes are nice, but hard evidence is better. Engineers do not build bridges from a left-wing or right-wing perspective. They build bridges from an evidence-based perspective and over time bridge-building gets better. Politicians make their decisions from an ideological perspective and over time our politics gets worse.

Be prepared to lose nine battles out of ten. You cannot win them all, but you will likely win some. In 'round two' be prepared to win one battle out of nine, and by round three one out of eight. By the time you've done that, you've created enough of a shift for others to follow. If you entered the fray worried about losing nine out of ten, you'd never start. Concentrate on winning the one. Overnight success is for the movies.

Kick out cynicism. Cynicism has become embedded in our society and is seen as wisdom. Yet there is nothing wise or even likeable about cynicism. Cynicism is a bit like smoking: you may think it looks cool but it's really bad for you – and worse than that, it's bad for everyone around you. For the cynic, everything is just a little too hard to imagine, or do. As such, cynicism is both a recipe and an excuse for laziness. Have no time for it.

This last principle is perhaps the hardest principle to practice, because cynicism is so seductive. How many times do I tell myself it's just not worth the effort, that I'm an idiot to even try and make a change? Daily. And then I remember Mohamed Nasheed.

As I write, President Nasheed of the Maldives has been deposed in a *coup d'état* – a 'win' for the cynics who said he was a hopeless dreamer. Many in the nation naturally fear a return to dictatorship, but that seems almost inconceivable, such is Nasheed's legacy in bringing democratic reform to the Maldivian archipelago. The ousted president retains massive popular support and is gaining ground in the battle to force a new round of elections so the people can decide again who should lead them.

I can still hear him, as he sat opposite me that balmy afternoon in Malé, saying 'There's always more than one avenue to any destination. Even in a dead end, when things get really, really bad you have to keep going – however badly you suffer, whatever losses you incur, you just have to keep going. You have to make a tiny step. Don't just be stuck with a single option.'

All the principles of pragmatic optimists like President Nasheed can perhaps be encapsulated in a single idea – one that I pondered on the steps of the New York Library late in 2010 – and the one I concluded my address to that gathering of the super-rich with.

I'd been flown to Budapest business class, put up in a super-posh hotel and asked to close a conference where the assembled wealth managers debated the future and specifically what to do with their billions of dollars of assets. As the economic crisis lunged from one precipice to another the mood was anything but optimistic.

'How's your week?' I asked one attendee over dinner.

'Pretty grim,' he replied. 'I'm down $213 million since Monday'. That's got to affect the taste of your steak.

On the podium I was way out of my comfort zone. I told them to lean into the curve of the future, to let go of the mindset of the Industrial Revolution, that they should invest in Tony Lovell's farms, carbon neutral fuels and new models of education. I said, 'if you're not actively trying to make the world better, what are you for?'

I told them the future was a game still worth playing and they had a role to play that they had no right to shun. 'We're no longer tenants on the planet, we're the landlord'. After forty-five minutes I concluded with my New York epiphany: 'Judge your worth not by what you *own*, but by what you *create*' ... and walked off stage expecting to suffer whatever it is the super-rich employ in place of a lynching. Death by hotel bath robe perhaps?

Instead some of them asked me to help them set up an investment fund with the stated principle of making the world better.

I'll probably get it wrong. Maybe we'll fail. But the list of principles I gleaned from those extraordinary people who filled the pages of *An Optimist's Tour of the Future* reminds me that making mistakes is OK, but not trying is irresponsible.

I was recently asked by an interviewer why futurology was 'back in vogue'. I'm not sure that it is – but if it is, then I'm pretty sure it's because we're all realising that the future is up for grabs again. We can see that many of the social systems and institutions we've been using are no longer fit for purpose, while at the same time technology and our advancing knowledge are bringing fresh tools and ways of thinking that can reshape our world.

Or to put it another way: we know we have to change the way we're doing things while at the same time the tools to help us do so are coming online.

So what *will* the world look like in fifty years, my interviewer continued.

My reply was that I try not to make too many predictions. The history of futurology tells you far more about the prevailing culture at the time, and/ or the particular prejudices and concerns of the individual futurologist than about what actually happened. I'm more interested in the questions technology poses us.

'What I will say,' I said, 'is that the future we have will be about the values we choose. We're at a point in history where it may be possible that those values won't be determined so much by physical constraints but by what we dream.'

I hope we have good dreams.

Mark Stevenson, Spring 2012

INDEX